EAT OR BE EATEN

Dr. RICK BEIN

WORKBOOK PRESS LLC
187 E Warm Springs Rd,
Suite B285 Las Vegas NV 89119 USA

Website: https://workbookpress.com/
Hotline: 1-888-818-4856
Email: admin@workbookpress.com

Ordering Information:

Quantity sales. Special discounts are available on quantity purchases by corporations, associations, and others. For details, contact the publisher at the address above.

Library of Congress Control Number:

ISBN-13:   978-1-96371-t801-0 Paperback Version

DATE: 9/26/2025

CONTENTS

INTRODUCTION

The issue of food can be interpreted in two ways. As a geographical agriculturalist, I have focused on food production around the world, but also with being faced with predatory action. These stories relate such events in my life. My Peace Corps experience provided the spark that led to this series of adventures and observations.

Farming strategies vary tremendously around the world, from my home farm in Colorado to those of Brazil, Sudan, Papua New Guinea, Mozambique, and Kenya. In these places there are also creatures, higher up on the food chain that come in to play.

The cover picture depicts a Sudanese feast, where various food items are laid on an outstretched table cloth on the ground to serve men dressed in formal attire. Only the men are eating and when they are sated the women come to eat what is left over. Notice the only the right hands are touching the food. The left hand is considered foul and would contaminate the food.

1. VALUE OF TRADITIONAL AGRICULTURE

The greatest challenge to understanding how traditional farmers maintain, preserve, and manage biodiversity is to recognize the complexity of their production systems. Today, it is widely accepted that Indigenous knowledge is a powerful resource in its own right and complementary to knowledge available from western scientific sources. Therefore, in studying such systems, it is not possible to separate the study of agricultural biodiversity from the study of the culture that nurtures it. (Altieri, M. A. 1987.)

Crop mix of dozens of plants and useful weeds. Photo by Rick Bein 2011.

Many plants within or around traditional cropping systems are wild or weedy relatives of crop plants. In fact, many farmers "sponsor" certain weeds in or around their fields that may have positive effects on soil and crops, or that serve as food, medicines, ceremonial items, teas, soil enrichment or pest repellents. In the Mexican Sierras, the Tarahumara Indians depend on edible weed seedlings or "quelites" (Amaranthus, Chenopodium, Brassica) in the early season from April through July, a critical period before crops mature from August through October. Weeds also serve as alternative food supplies in seasons when maize or other crops are destroyed by frequent hailstorms (Bye, 1981). In barley fields, it is common for Tlaxcala farmers to maintain Solanum mozinianum to levels up to 4,500 plants/hectare, yielding about 1,300 kilograms of fruit, a meaningful input to agricultural subsistence (Altieri and Trujillo, 1987).

As more research is conducted, many of the traditional farming practices once regarded as primitive or misguided, are being recognized as sophisticated and appropriate. Confronted with specific problems of slope, flooding, drought, pests, diseases, low soil fertility, small farmers throughout the world have developed unique management systems aimed at overcoming these constraints (Klee, 1980). In general, traditional agriculturalists have met the environmental requirements of their food-producing systems by concentrating on a few principal characteristics and processes resulting in a myriad of agricultural systems that store the following structural and functional commonalities (Gliessman, 1998; Altieri and Anderson, 1986).

In contrast with commercial agriculture, traditional agriculture can rarely contribute to monetary wealth but can successfully provide a healthy subsistence for farmers. However unadulterated plants from traditional agriculture can be gleaned to strengthen the genetics of commercial crops.

2. BIO-DIVERSITY INVENTORY OF KAMIALI WILDLIFE MANAGEMENT AREA

THE BIO-DIVERSITY INVENTORY: CASE STUDY OF KAMIALI WILDLIFE MANAGEMENT AREA

F.L. (Rick Bein

Environmental Research & Management Centre

PNG University of Technology

Lae Papua New Guinea

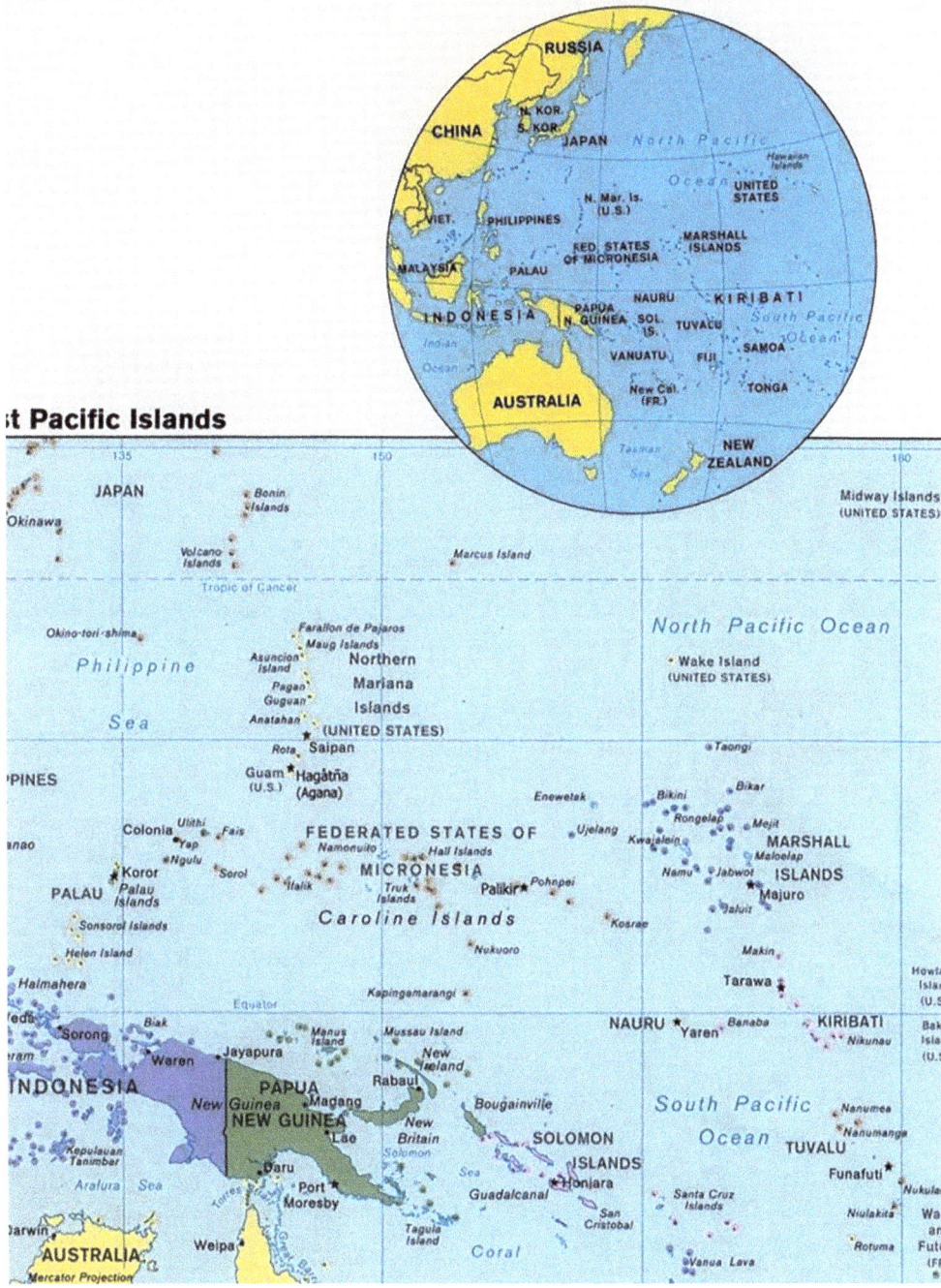

st Pacific Islands

Location of Papua New Guinea 0_____500 miles **Map for the Village Development Trust by CIA.**

Abstract

A team of scientists in 1997-98 inventoried the biodiversity of the Kamiali Wild Life Management Area, 60 kilometers southeast along the Huon Gulf coast from Lae, Papua New Guinea. The 434 square kilometers area was registered with the PNG government as a wildlife management area in 1996. A village of 500 people has elected a committee who with the Village Development Trust (VDT), a nationally based NGO, manage the area extending 12 kilometers out to sea and 17 kilometers inland. The high relief environment ranges from 1080 meters below the sea up to 2012 meters above mean sea level. A transect method was used to sample biomes beginning with open sea, coral reef, mangrove, beach, village, sago swamp, garden, riparian environments, tropical forest and cloud forest. This study provides a baseline inventory for a unique area which has never experienced hillside gardening nor commercial logging and contains a high biomass environment that serves as a habitat for many rare species. This case study demonstrates an application for the conservation of natural "Resources for Science and Technology in Development".

The Role of Conservation and the Wildlife Management Area

"Conservation of resources" is a concept which lies half way on the continuum between "exploitation of resources" and "preservation of resources". Resources exploitation is the extraction of nature in a manner in which there is no regard for the long term future benefits that those resources may provide for humans. The antithesis of exploitation is the preservation of resources which in the purest sense promotes the total non-use of the resources by humans. Neither of these extreme approaches can functionally serve human kind for the long term. Humans must use the earth's resources in order to survive and conversely, humans will perish if they consume all of the resources at once. Conservation on the other hand promotes the sustainable use of natural resources.

Wildlife Management Areas (WMA's) have been established for years in a number of countries (including Papua New Guinea) to slow down the process of exploitation of natural resources. It falls slightly left of center on the preservation side of the continuum. Another concept, the Integrated Conservation and Development Area (ICAD) is a recent addition to the many methods of protecting the natural environment from human destruction. It promotes sustainable human use of the resource and occupies a more central position. In PNG, WMA's have been more commonly established because there are government regulations to follow. Regulations for the implementation of ICADs did not exist in PNG until recently. In many situations the WMA's of PNG are now including the thinking behind the ICADs by incorporating the concept of "sustainable use of resources". This is the current position of the Kamiali Wildlife Management Area (KWMA).

"Conservation" is quite different from "preservation" which promotes total disuse of the environment by humans. Although the idea of preservation has merit in small isolated parts of the earth, it is largely impractical with growing populations and their increasing demands on the natural resources. Conservation on the other hand takes the position, "That if humans are to survive on this planet they must strive to find sustainable uses of the environment". Should people fail to do this they will perish. The planet has been around for a very long time and it will continue to be around for a very long time. The question is: "Will people continue to be along for the ride?"

Conservation of biodiversity

Biological diversity is the richness of our planetary ecosystem and includes genetic diversity, species diversity and ecological diversity.[1] The conservation of biodiversity is important not only to ensure survival of millions of species but also to support the living systems of the world's people who depend on natural resources and the ecosystem. Leading ecological thinking considers high bio-diversity to be a desirable thing for the planetary ecosystem. Following the ecological premise that every living thing has a "niche" or functional purpose in serving the greater ecosystem, it would hold that by maximizing the number of living species, there would be more opportunities to maintain ecological balances. As a part of the global ecosystem, humans would benefit the most from high biodiversity. Benefits found in nature take the form of foods, medicine, building and clothing materials and chemical substances. Many discoveries have benefited human kind and have over the millennia fostered the expansion of the population. Maintaining bio-diversity provides continuous human benefit. Preserving bio-diversity does not contradict conservation; conservation when operating properly preserves bio-diversity.

It further happens that tropical biomes such as those in PNG contain some of the highest bio-diversity on the planet. Hundreds of species, many yet to be named, thrive in the PNG rain forests and coral reefs. The world is eager to discover these new species, their ecological niches and the benefit they can provide human kind. PNG is still 60% forested[2] much of which is pristine and contains some of the richest biodiversity in the world. There is great fear that PNG will lose this resource if current trends of deforestation continue.

Dichotomy between economic development and Conservation

Papua New Guinea, on the other hand, is striving to develop economically and coupled with a 30 year population doubling rate it must exploit its basic resources. Traditional paradigms and lifestyles of the village are changing. Attitude toward the environment is also changing. Western paradigms, promoting the monetary economy begin to erode the old lifestyles, and the environment becomes expendable for the purpose of seeking immediate wealth. During a paradigm shift the natural environment becomes vulnerable and easily exploited to support the new way of thinking. Exploitation occurs during the transition when old thinking gives way to, or merges with new thinking.

Kamiali

Kamiali view of Blue Mountain and VDT housing. Photo by Rick Bein 1998.

Application to Kamiali

Kamiali became a WMA because of several unique circumstances:

1) It is the only coastal village along the Huon Gulf between Lae and Oro Province which chose not to sell their trees to the Malaysian logging companies during the early 1990's and

2) The food garden soils in the fertile Bitoi River flood plain provide enough food that there has been no need to clear the highly erodible hill sides.

As a result of these two factors, much of the forest remains pristine, and the coral reefs fringing the coast of Kamiali are equally pristine because they are free from the smothering effect of sediments eroding from the land.

Since Kamiali is mainly a fishing village, there is very little hunting pressure and the original wildlife is intact. The three village hunters have had little problem finding wild game with in short distances of the village and as a result rarely venture up into the remote areas generally above 300 meters elevation.

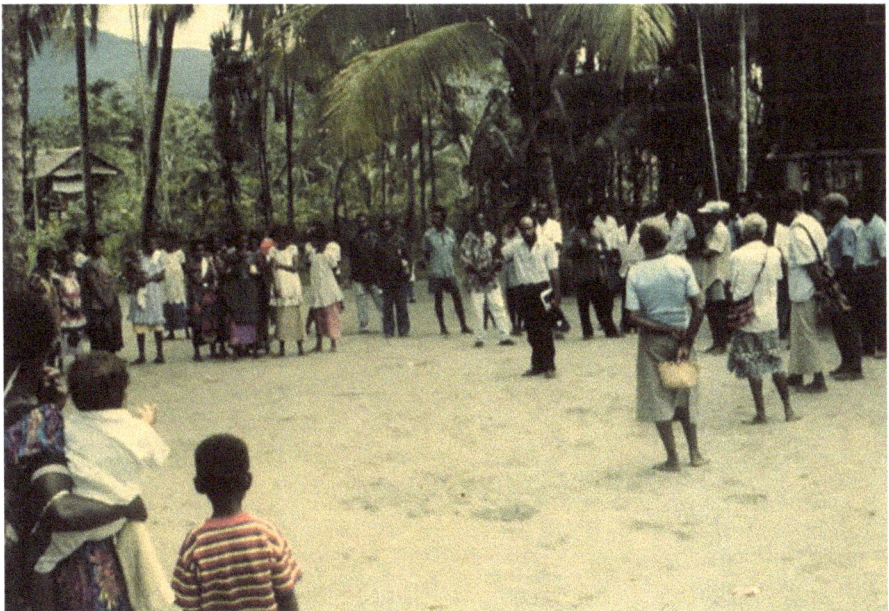

Village meeting: VDT representative addresses the Kamiali villagers.

Methodology and Logistics

The biodiversity inventory was conducted during the period between July 1997 and July 1998. When the Lababia community established the Kamiali Wildlife Management Area, the Village Development Trust of Lae PNG, obtained a grant from the New Zealand High Commission to conduct this study. The grant provided for the listing of species found in the various biomes through out the Wildlife Management Area and the listing recommendations regarding further study and better conservation of the wildlife management area. The sampling technique comprised a transect from the open sea through the coral reef, mangrove swamp, beach, freshwater mangrove, sago swamp, lowland rainforest, lowland hill forest, upland and montane cloud and moss forest. Also sampled were sites that the scientists deemed to be notably different such as Lababia Island and the Bitoi food gardens. No attempt was made to numerically quantify any part of biodiversity except in the case of the fish study. Nor was there any attempt to sample the genetic diversity.

A signed "scientist, good faith, research agreement" has been established to provide incentive for the villagers to host scientific research and to protect their intellectual property rights. This includes an understanding that participating scientists:

1) build into their research proposals, cash grants as contributions to the community,

2) employ villagers when ever possible in the data collection and educate them within their desire to understand about the research and its methodology in order to strengthen their abilities and enthusiasm to support future research, and

3) commit to share the proceeds, should the scientist or representative institution make a profit from the discoveries at Kamiali, as a continuing contribution designated to support village causes such as the payment of school fees or for the purchase environmental conservation technology.

Facilities to accommodate field research scientists include the comforts of a guest house with prepared meals, and a training center built by the Village Development Trust, air compressor to refill diving tanks, storage capability, easy access to most biome areas, transect trails to the more remote environments, and boat transportation from Lae. Up to 16 people can be accommodated in four person screened bunk houses and another 20 people in a dormitory bunk room. Running water supplies clean showers and toilets.

Preliminary results

Thirteen participating scientists focusing on different parts of the biota have found many unique aspects of the bio-diversity of the Kamiali area. Briefly, the high lights of the findings show that:

1) the fringing coral reef is one of the highest biomass reefs in Papua New Guinea,

2) the undisturbed lowland hill forests next to the ocean shoreline are unique because they have not been periodically cleared for food gardening like most PNG coastal communities,

3) vegetation typical of 2000 meters elevation is found as low as 600 meters on ridge slopes facing toward the sea,

4) a vast area of land has not been untouched by humans for at least the last 50 years,

5) Lababia Island serves a major roosting site for frigate birds and pied imperial pigeons,

6) the broad sand beaches of Kamiali are major nesting sites for leather back turtles, and

7) several undescribed insect species are in the process of being recorded.

The complete inventory contains the appended reports of each scientist including all species found and those species, not seen, but which are probably there because of their association with other known species. Also included are recommendations made by the scientists for maintaining the biodiversity and for future research.

Arial view of Kamiali Village Photo from Village Development Trust collection.

A base for further research

The biodiversity inventory at Kamiali with its logistical facilities in place provides an excellent basis for further research. The potential for future research in such a pristine environment is unending. The major difficulty is the long distance travel and time required to come to Kamiali from the research institutions around the world.

This inventory is not all inclusive as the sampling technique used could only probe a small area. There are great research possibilities in the further sampling of the biomes. The cloud forest biome has been barely examined for plants and insects. The coral reefs and mangroves and beach areas offer many opportunities. Over 100 leather back turtles nested in Kamiali this last season and discussions are underway to experiment by incubating some of the eggs to ensure greater hatch rates.

[1]Miller, G. T. Jr. (1996) "Living in the Environment" 9th ed. Belmont, California, USA: Wadsworth Publishing.

[2]Filer, C. and Sekhran, N. (1998) "Loggers, Donors and Resource Owners". Boroko, Papua New Guinea: The National Research Institute.

3. OUR CROPS IN BERTHOUD

We managed a mixed crop and livestock 200-acre irrigated farm just east of the town of Berthoud, Colorado. We grew sugar beets, barley, alfalfa, corn, pinto beans, and cucumbers. Corn supported the beef fattening program and a small family milking practice.

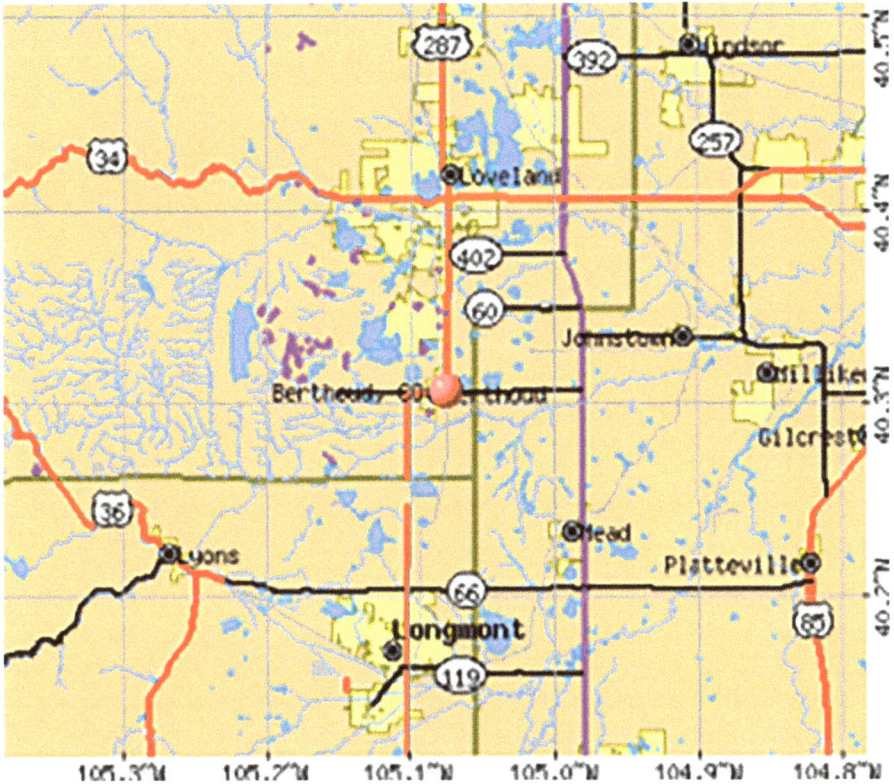

Our farm was located two mile east of Berthoud, found in the center of this map about half way between Loveland and Longmont. The Rocky Mountain rise starts on the western edge of the map. This Map is copied from the Colorado State Highway map from 1960. 0____3 miles

Our feedlot operation had a capacity of 1000 bovines. Steer and heifer calves, recently weaned, were purchased from cattle ranchers in Wyoming and trucked to our feed lots. The six-month-old calves missed their mothers and together would create a mooing chorus that continued for about a week.

Our cattle feed lot with our farmhouse in the back ground. Photo by Rick Bein 1960.

We fed them a mixture of ground corn and corn silage and alfalfa. This diet varied as we substituted green chopped alfalfa in place of corn silage during the Summer, and in late Fall, the cattle were allowed to graze the harvested sugar beet field for the discarded beet tops. When we switched to grazing the harvested beet field, cattle occasionally clogged their throats on extra-large beet tops. We had a few animals die from this. When chopped alfalfa replace the corn silage in the summer the cattle had to be watched for bloating.

The stomach would become quite distended when an animal bloated, We made the steer move and walk around to allow the gas to escape normally. In severe cases a sharp hollow blade was stabbed into the bulging stomach to relieve the gas before the cow died of heart failure.

When a cow had a beet top caught in its throat, it could not eat or drink. We would push a thin garden hose down the throat of the cow to force the beet top down into the stomach.

A diet of fiber, like hay, grass, and chopped greens was normal for cattle and after three years the cows would be ready for market. We added corn to fatten them in two years. The ground corn was a treat for the cattle but occasionally one steer, would selectively eat only the grain and a condition called "foundering" (laminitis) occurred. It could kill the animal or leave it with locked-up, stiffened leg joints. We had to avoid this by not overloading the diet with too much grain.

We rotated our crops every year except for alfalfa. Alfalfa is a perennial that grows back every year and has a head start every spring because it does not need to be replanted. However, each succeeding year the productivity declined and after the third year it was plowed under. Usually, we would plant sugar beets on that land the following year. Alfalfa is a hardy legume that fixes nitrogen in the soil that benefits the next crop in the rotation.

Alfalfa grows quickly and the three cuttings per season provided us with enough hay (baled or stacked loosely), to feed the cattle. We also chopped up alfalfa as green fodder. We made alfalfa silage out of this green "chop" to provide fodder during the winter months.

Silage was made by storing chopped up plant material, normally green corn, but also green alfalfa for several months while fermentation occurred. This pickling effect created an alcoholic content of about 1% which the cattle enjoyed. Initially we used upright silos (still dotting the agricultural landscape of the USA) for the silage producing process. Vertical elevators were used to fill the silos and gravity was used to empty them. They did not take up much ground space and the silage was easily tossed down to be distributed to the near by cattle. Later we used horizontal trench silos carved into hillsides where it would be removed by mobile loading equipment and hauled to the feedlots by trucks. It was fun however to climb up the vertical tower silos for a better view. But in the 60s safety regulation was established that required the ladders to be removed when not in use.

.

Our farm yard was rimmed by cottonwood trees and the distant haze of the Rocky Mountains.

Sugar beets played a mixed role on the farm. The fifteen-pound beets were dug out of the ground and immediately trucked to the Great Western Sugar factory in the town of Loveland, seven mile to the north. The beets were crushed, and the sugar juice processed into the granular sugar. The leftover wet beet pulp was loaded onto the same truck and the farmer took home another form of cattle fodder. This pulp still had a sweet taste, and the cattle loved it. The sugar factory was unable to dispose of all the wet pulp at harvest time, so they made dried beet pulp available to feed cattle later in the year.

Mechanized sugar beet harvester loading beets into a truck in November. Photo by Jean H. Bein 1963

Sugar beets were introduced to Colorado in the early 1880's as an alternative to sugar cane by Ukrainian immigrants. Sugar cane can only be grown in the tropics and is much cheaper to produce. In order to guarantee a ready source of sugar the US government instituted a crop subsidy for farms growing sugar beets. Sugar beets originally came from northern Europe. The beets grow well in colder climates where pests are less of a problem. Farmers applied for an allotment with the sugar factory as to how many acres of beets that could be grown. We considered the sugar beets our main cash crop and were allowed to grow 30 acres.

Sugar beet growing took a lot of hand labor. At that time, the germination rate of the sugar beet seeds was not that high and planted them in clusters and when the plants emerged, they grew too close together to allow a fifteen-pound beet to grow. Thinning with hoes had to take place. It was done by braceros who came from Mexico each summer to perform farm labor tasks. We provided a small house where they stayed. As a small boy, I used to hang out at their place, and they taught me some Spanish.

We grew barley for the Adolf Coors Brewery in Golden, Colorado and we contracted with the brewery to grow 20 acres. The barley was stored in our granary until the Coors truck came. We always kept some Coors beer around for refreshment. Until the 60s and 70s, beer was not allowed to be marketed beyond state lines. Coors was a popular beer and people came from surrounding states to buy it. When we traveled out of state, we would carry a case as a gift for our friends and family. When I returned from the Peace Corps, they had developed something called "Coors Light." I could never see the reason for this because the original was already "light"! But then, the original Coors beer did not taste the same.

There were two minor crops we grew for a few years: pinto beans and cucumbers. Pinto beans were grown to remedy a shortage in Mexico and were immediately exported after harvest.

Loose pinto beans and canned pickles in jars. http://www.photos-public-domain.com/2012/04/23/pinto-beans/ Canned Dill-PIckles-1280-1524-1

We sold the cucumbers to the pickle factory in Loveland. It was fascinating to watch the conveyor with the slots that sorted the pickles by size. They immediately fell into a brine vat where they remained until canning. Back home, Mom and sister Jeanie canned enough dill pickles to provide for several years.

Hybrid corn was a major crop as it was critical food for the livestock. As I mentioned before we fed the cattle with ground shell corn and corn sllage. Hybrid corn involves cross breeding by controlling the pollination process. Two different varieties of corn are planted side by side in a field, but the pollinating tassels are removed from one of the varieties. The variety without its tassels is pollinated by the tassels of the other variety. This process was repeated again with the results of another cross-pollinated pairing. Seeds of this second crossing exceeded the productivity of the normal open-pollinated corn by 2 to 3 times. This is what we called "hybrid vigor. Unfortunately, the hybrid vigor seed was sterile and did not germinate when planted. At that time in the 1950s it was the state of the art of corn farming. Since that time, science has taken this hybridization process several steps beyond to produce genetically modified corn (GMO). Each year the seed has to be purchased from the corn breeder.

Field corn was strictly used for animal feed and so we grew sweet corn for our household meals. Sweet corn was harvested mid-summer when the seeds were still soft and juicy. Mom and my sister Jean stored some in the deep freezer.

I had an intimate experience contributing to one of the hybrid corn crossings. My Dad's friend, who was a corn seed producer asked me, as a fourteen year-old to monitor the initial cross breeding of two varieties. A one-acre plot isolated a half mile from any foreign pollen coming from other corn fields was set aside next to our house. My job was to detassel one of the varieties by repeatedly going through the rows of corn. I did such an excellent job that the seed producer paid me $2000.00, which was a lot of money in those days.

I participated in other projects. Our two milk cows needed to be milked by hand twice a day and I was the logical candidate to do this chore. Learning to extract the milk from the teats was a challenge which soon became second nature. I made it fun by using the teats as squirt guns by shooting streams of milk at the cats and anyone who came to watch the milking. The cats would rise up on their back feet catching the streams in their mouths. Visitors were surprised with an eye full of warm milk when they entered the milking parlor. The two cows produced more milk than we needed so dad invited the tenant farmer to do the evening milking allowing him to keep the milk. We made many products from the raw milk including cream, butter, butter milk, skim milk, cheese and cottage cheese and of course whole milk!

I was allowed to raise my own animals. As a 4H project I entered the catch-it-calf contest at the National Western Stock show in Denver. I caught a calf and brought it home and a year later brought the full-grown steer back to the Stock Show where it took third place for weight gain.

In my sophomore year in high school, one of the local meat packing plants wanted to conduct an educational project for our 4H club. They gave each of us (ten girls and boys, aged 9 to 14) a calf to fatten all the way to the grocery store. We each kept detailed records of the amounts feed, the costs, and the monthly weight gain. At the end, we brought all of our records and grown steers to the meat packing company. They praised us for all our work and gave awards for the best bookkeeping, interesting practices, and most weight gain. We said goodbye to our steers and were told to come back the next day. We did not witness the slaughtering, but they showed us our animal carcasses hanging from a rail. The packers proudly praised our participation in this great this learning experience. As each carcass was identified to its owner, we were shown which parts were the best cuts of meat, the whole experiment fell apart. Weeping and wailing took over! Kids were traumatized, that something like this would happen to their best friend! One nine-year-old girl never got over it! The packing plant did not repeat this project.

One year, dad came across small flock of lambs to fatten. I took over the care of them and I got to keep a share of the profit. From time to time, we kept an orphaned lamb running around loose in the farmyard which my sisters took over the bottle feeding.

For my high school Future Farmers of America project, I was given a pregnant gilt for which I had to give back two female piglets of the first litter, so two other kids could do the same thing. Beginning with that sow, I had amassed one hundred pigs by the time I graduated from high school. The sale of the whole lot financed my first year in college.

VMy "Yorkshire White" pigs. Photos by Rick Bein 1959.

Farm life served me well. In addition to all the experiential knowledge I gained, I qualified to serve as an agricultural extension agent in the Peace Corps.

4. WOLVERINES

An Encounter with Wolverines

In the Summer of 1964, I was working for the Rocky Mountain National Park Service. Based in Grand Lake, Colorado on the western slope of the Rockies, we were about 15 young men, average age of 20, performing trail maintenance tasks. We were broken up into small groups to work on different projects. A fellow from Greeley and I were assigned to repair a hiking trail up in the tundra at an elevation of about 10,000 feet.

Tundra of Rocky Mountain National Park. Photo by Rick Bein 2010

The trail that we were working connected Grand Lake with Estes Park crossing over the Continental Divide between 14,000-foot mountains and over to the Eastern Slope. We were working the western slope of the trail. Because the slope was steep, a series of "S" shaped switch backs made it easier for hikers to move up and down. The pristine environment was only disturbed by this trail.

One day in late July the two of us were working a series of switch backs, clearing rocks and stabilizing places where the trail had been eroded. My partner was two switch backs above me when two wolverines began crossing laterally up slope above him. They were loping along, not a care in the world, hardly noticing us. These fearless creatures known to attack animals many times their size, reportedly killed a polar bear in a Minneapolis zoo enclosure. They are said to have keen sense of smell.

Wolverine by Skansen an existential hero is licensed under CC BY-NC 2.0

He thought it would be clever throw a few stones at them. The two wolverines stopped and looked at him as if to say, "Who are you to be harassing us?" and then started straight down the slope toward him! Terrified, he panicked, dropped his shovel (his only weapon) and raced down toward me. I tackled him to keep him from falling over the cliff behind us, and then I turned to face the wolverines!

They were still coming!

I too was scared but I knew running would make things worse as they could run faster than I! Instinctively remembering what my father had taught me when warding off an angry bull, I raised my outspread arms over head while brandished my only weapon, the shovel!

For some reason they stopped about 10 feet away and looked at us and then decided this was not an encounter they wanted! They then continued their leisurely trot off to where they were originally headed. What a relief!

Then I noticed a terrible stench. I recognized that smell and when I looked around at my partner behind me, I knew it was not coming from the wolverines! Maybe that was why the wolverines stopped coming and it had nothing to do with waving my arms!

Now you know what to do the next time you meet some wolverines!

5. PIRANHA

Men should Never Swim with Piranha without a Swimming Suit On!

Piranhas in South American rivers perform a job for the environment by keeping the rivers clean of organic material. While known for their meat-eating habits, not all of the dozen or more species of piranha eat flesh. Some species of piranhas are vegetarian and consume fruits and leaves that fall into the rivers.

My Peace Corps (1964) experience with piranhas came after an extensive all-day trip from the Campo Cerrado down onto the Pantanal of western Brazil. A settler and his family were being taken out to the Taquari River to where they would be sequestered once the rainy season was in full force. My Peace Corps buddy, Richard Bartles, and I accepted the offer and joined the group.

CENTRAL MATO GROSSO

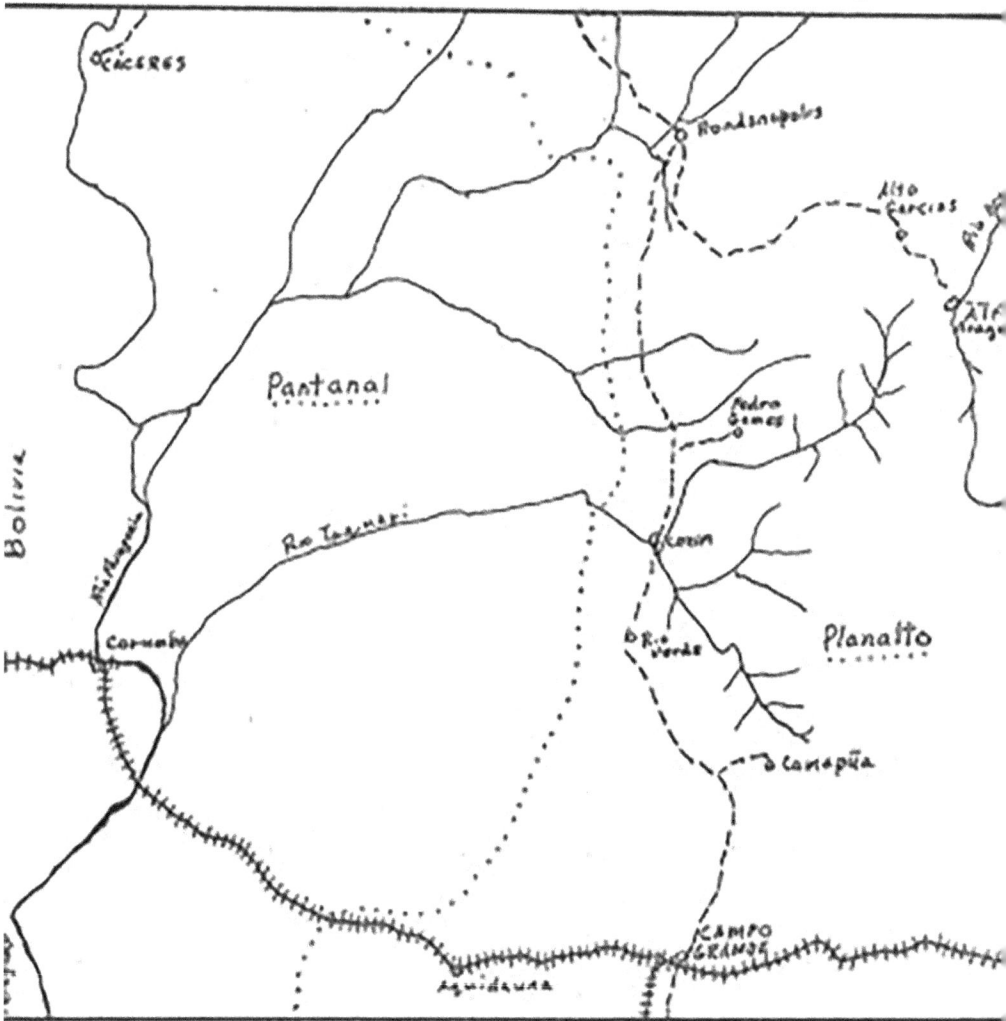

Scale 1: 3,500,000

Hand drawn map by Rick Bein 1966 showing the towns of central Mato Grosso do Sul with the Pantanal on the left side.

0_____200 miles

The Pantanal of western Brazil is one of the world's largest unique wetlands. Most of the seventy-thousand-square mile area (about the size of North Dakota) is owned by approximately thirty ranchers who use it mostly for dry season cattle grazing and recreation. During the rainy season, this vast flat lowland floods from all the surrounding swollen rivers. It is only one hundred meters above sea level, yet it is two thousand kilometers to the sea. The Paraguay River slowly drains the Pantanal each year, taking the water south to meet the Paraná River, which releases into the La Plata River before dumping into the Atlantic Ocean.

The Pantanal has one of the most unique bio-diversities. A number of swamp plants have evolved very differently here from similar environments around the world. Among the fauna are a number of unique species pictured below. Some are shared with the Amazon Basin such as the Piranha. The largest rodent on the earth, the capybara is found here, a flightless bird unique to Brazil, the ema, the peccary, the anaconda and many more.

The Pantanal – Passeo de Longes Camp – Capybaras (Hydrochoerus hydrochaeris) , largest rodent on the planet. Photo by kthomason is licensed under CC BY-NC-ND 2.0.

Ema (Rhea Americana) by Wagner Machado Carlos Lemes is licensed free under CC BY 2.0

Collared Peccary (Pecari tajacu) by Smithsonian's National Zoo is licensed free under CC BY-NC-ND 2.0 2008

Anaconda_Loreto_Peru httpsupload.wikimedia.orgwikipediacommons2009

It was the beginning of the rainy season and the last opportunity to move the family before the Pantanal became completely flooded. Half a dozen of us crowded in among all of their furniture plus the food and supplies that were loaded onto the open truck.

After leaving the main road and starting down onto the Pantanal, someone discovered the case of Caxasa bottles hidden under some of the family's belongings. A lid came off, and the bottled alcohol was passed around. Eventually some bottles ended up in the truck cab! A short time later the driver became inebriated and several times ran the truck into the mud! Each time everyone climbed down and helped push the truck back on the road.

Caxasa (pronounced Kashasa) is distilled from sugar cane resulting in a cheap but strong alcohol available throughout Brazil. Decades ago it was discovered that it could also be used as an emergency fuel for automobiles! Since then it has been refined and used as vehicle fuel to avert the world petroleum crises that began in the 1980's. Today Brazil is petroleum independent and arguably the world leader in ethanol production.

About an hour before dusk, we finally arrived at the ranch house on a levee overlooking the Taquari. We were sweaty and filthy from the trip, and the horseshoe shaped waterbody fifty meters upstream was inviting. We joined a group of Brazilian men bathing in this small oxbow off the river. Wading out waist-deep, we enjoyed lathering up and leisurely rinsing off. We did not pay attention to the Brazilians, who quickly got wet, jumped out to soap up and then quickly returned to rinsed off, and immediately leave the water.

When we finally waded out of the water, a cabocolo with a high-pitched voice came by and said in Portuguese, "Voces estao com muito sorte, homens nunca nada com piranha sem calsao"! (You were lucky this time ... men should never swim with piranha without a swimming suit!)

After that sobering remark, we returned to the ranch house. The evening was one to remember as we dined on a freshly killed calf and listened to an hour of piranha stories.

One of the stories follows. Just before the rainy season when the Pantanal would begin flooding, the ranchers would herd their cattle out of the swamp, up onto the Central Brazilian Plateau. Thousands of these cattle were herded a thousand kilometers over to Sao Paulo, where they would be slaughtered to provide meat for the people living in the large urban areas. But along the way they had to drive the cattle across piranha-infested rivers. The piranha would take bites out of several cows, and these bites would become infected down the road and some of the cows would die. To avoid this, before crossing, the ranchers selected their most sickly cow, pushed it into the water, shoot if full of bullet holes, and let it float down the river for two hundred yards. The piranha feasted, and when their bellies were full, the rest of the cattle crossed unmolested.

I retired for the night but had a hard time sleeping. The mosquitoes were swarming and I was having awful nightmares huddled with my knees tightly together under the bed sheet, the closest thing I had to a mosquito net.

Early the next morning, several of us went fishing for piranha in the very spot we had taken our baths. In less than a half hour, we caught two dozen piranha. Raw calf's liver, left over from the previous night attracted the piranha, who instantly stole the bait off the hook!

The trick to catching a piranha was to cast the baited line out into the water, count to three, and yank. That usually hooked a piranha, but I quickly found out that as I reeled the fish in, only the head of the fish remained on the line. To prevent the other piranha from eating their injured colleague, the hooked piranha had to be quickly jerked all the way onto the land.

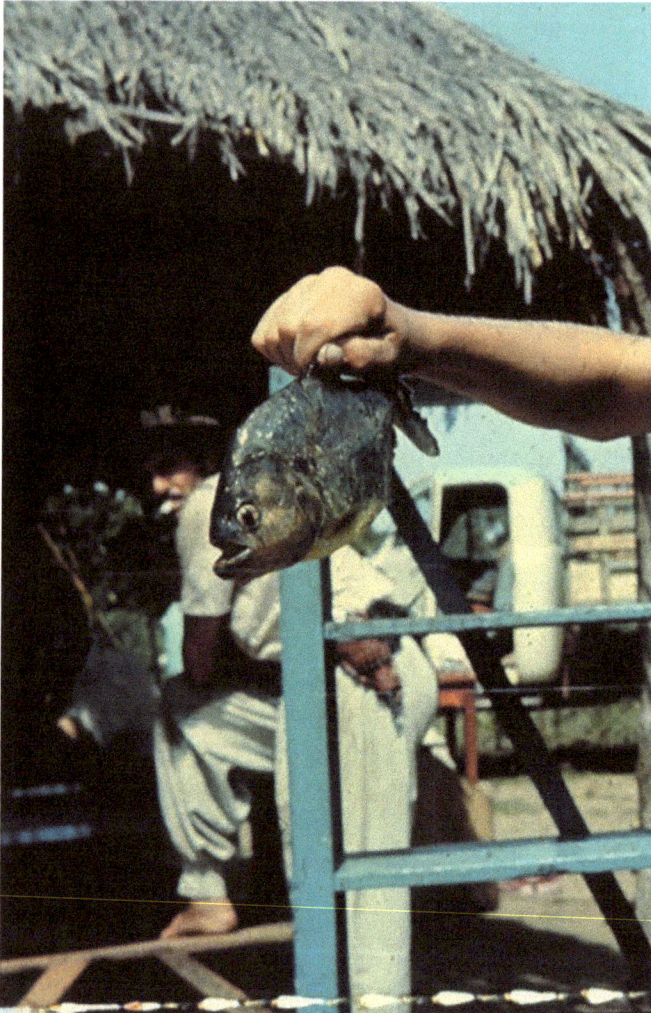

I don't know who was more fearsome the guy behind with his pistol and corn-husk cigarette or one of the Piranhas I caught. Photo by Rick Bein 1964.

Nor was it easy getting the hook out of the piranha's mouth. When I dangled the fish on the wire leader and hook, the piranha in its fury was still able to lunge upward and snap at my hand. I was terrified. I found out the locals left the hooks in the mouths until after the fish were cleaned, cooked, and served. The centerpiece of lunch table was a bowl where the hooks were deposited.

Are piranhas good to eat? Richard, my Peace Corps colleague, thought they were good. I thought they were tough and bony with a fishy smell and taste. Because of all the tiny bones they were fried to a crisp giving them a pretzel like crunchy texture. All I can say is, it was just as well that I was eating them rather than the other way around!

Thirty years later, my six year old son heard this story and said, "Dad, you were really lucky when you were swimming in that water!"

"No, Alex," I said, "*You* were the lucky one!"

6. ANACONDA

Anaconda from the Municipio do Pedro Gomes. Mato Grosso, Brazil

As an agricultural extentionist in my Peace Corps assignment I spent much time visiting farmers around the local county. This visit was one of the more interesting.

Twenty foot long anaconda was killed in their pig pen in 1962. Photo by Rick Bein (Third from the left) 1965.

This anaconda (Eunectes murinus) (local name: Sucuri) skin is what remained of the great snake harvested by these Brazilian pioneer farmers who had settled in Mato Grosso (now Mato Grosso do Sul) in the late 1950s. This family migrated from the Brazilian State of Rio Grande do Sul, to create a new life for themselves in the vast "unclaimed" wilderness of the headwaters of the Taquari River, 40 kilometers upstream from the renowned Pantanal swamp. All eight of their children had been born in Rio Grande do Sul and now the grandchildren were native to Mato Grosso.

There was a clearing on the south bank of the Taquari River where they decided to build their homestead, not knowing how the clearing came to be in the first place. They began clearing forest land to grow upland rice to provide sustenance for the family. After successive crops, the land became infested with weeds. New land was cleared and they introduced some exotic African grasses on the old weedy land. The Jaragua and Colonial grasses took over to form excellent pastures for the zebu cattle. On the following map the farm is located just west of the town of Pedro Gomes in the top center.

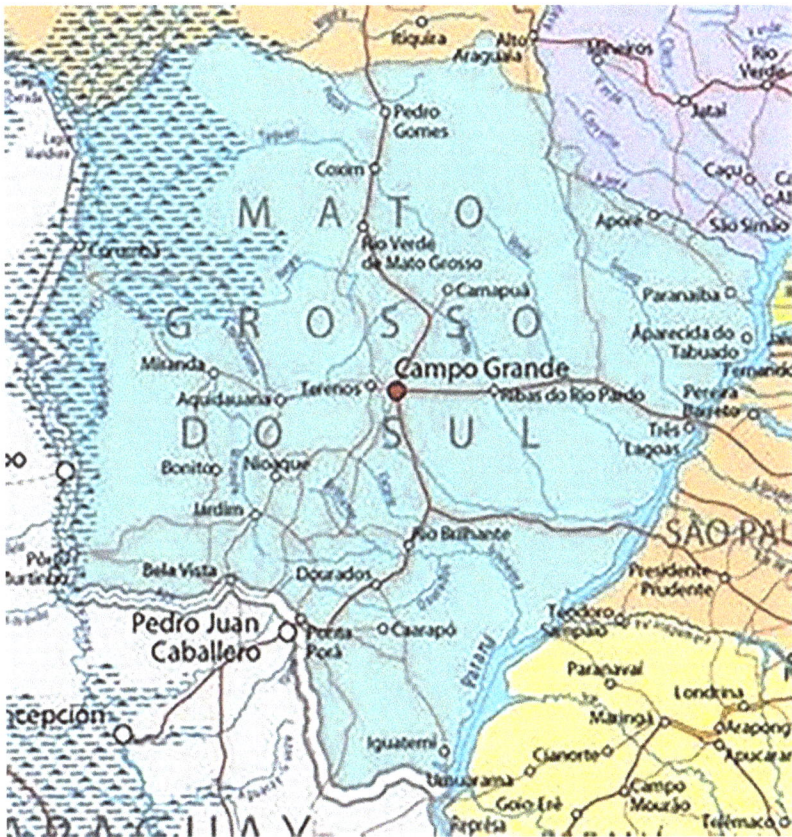

Pedro Gomes town sits in the north central part of the State of Mato Grosso do Sul. Free map from Maphill. 2011

0——————60 miles.

The family was terrified one morning when they came upon this 20 foot monster that had swallowed one of their pigs! It was total chaos as they battled the green anaconda (Eunectes murinus). The snake was unable to escape the pig pen because it now had a bulging belly with a pig inside, too large to pass through the fence rails. Of course it could have passed over the top of the fence, but that had not occurred to the snake.

Fearfully the family ganged up on the anaconda with spears and clubs as he was now fighting for his life. It was too dangerous to get close. After an hour of shooting at it with pistols and rifles the snake was finally killed. The family was quite proud of themselves and were eager to tell their story.

The anaconda had been living in the river next to the original clearing by their home for many years before this fateful day. No one dreamed that the snake was living under the water. They tied their canoes and the children collected water there. What they did not know, was that the anaconda was harvesting the collard peccaries (Pecari tajacu) (wild pigs, locally called javelina) that came in large bands initially over one hundred, for their nightly drink at the river. The original clearing along the river had been beaten open by the peccary's nightly pilgrimage. Once a week the snake would reach out of the water, over the crowd of peccaries, hardly noticed, and pluck a peccary out of the bunch. Immediately, it submerged, constricting and drowning the javelina before swallowing it. Meanwhile, the one peccary was hardly missed and the horde returned like clockwork the following night. Thus the domestic pig was not a major change in the snake's diet. The skin of a peccary is being held in the first picture by me and the oldest son dressed in the red shirt.

Collared Peccary (Pecari tajacu) by Smithsonian's National Zoo is licensed under CC BY-NC-ND 2.0 2008

The peccaries were considered pests by the farmers because they were destructive and dangerous as a large band of them could knock down a person and devour them in a few minutes. The peccaries disliked the structures that the humans had built on their trail and at night would break through the walls wreaking havoc to the home. To remain safe, the family would sleep suspended above the floor in hammocks and shoot blindly with a shot gun onto the dirt floor hoping to kill one of the pillaging pigs. Little by little they eliminated the peccaries.

Meanwhile the anaconda began to go hungry. Its next choice of food was critical, Fifty meters down the river the farmers had built a domestic pig pen with a corner extending into the shallow water of the river. This avoided having to carry water to the pigs. One evening the anaconda decided that a pig would be a nice alternative to a peccary; after all, they were in the same family even though a bit larger.

The farmers used most parts of the snake and had not found use for the skin. The father offered it to me for the small price of ten dollars. I bought the twenty footer and used it to show off to different audiences including my university classes. After 50 years with the skin someone offered to cure the hide and made it is flexible enough to unroll in my classroom. However, rolling and unrolling the skin took its toll as scales would fall off. After I retired I did not have many audiences to show the snake and decided to find a new home for it where it could be used for educational purposes. The White Pine Wilderness Academy for after-school children in Broad Ripple, Indianapolis agreed to mount the skin permanently on a wall where people can view it.

7. FLEES IN COXIM

An Adventure with Fleas

I had a 12 hour layover in Coxim (Pronounced Kosheem), Mato Grosso, Brazil while waiting for the Cuiaba bus. Coxim was the administrative center which includes the town of Pedro Gomes where I was serving in the Peace Corps. The only bus out of Pedro Gomes left at six in the morning and took two hours to arrive in Coxim at eight AM. Although it was 35 miles away, it took that amount of time to traverse the potholed gravel road.

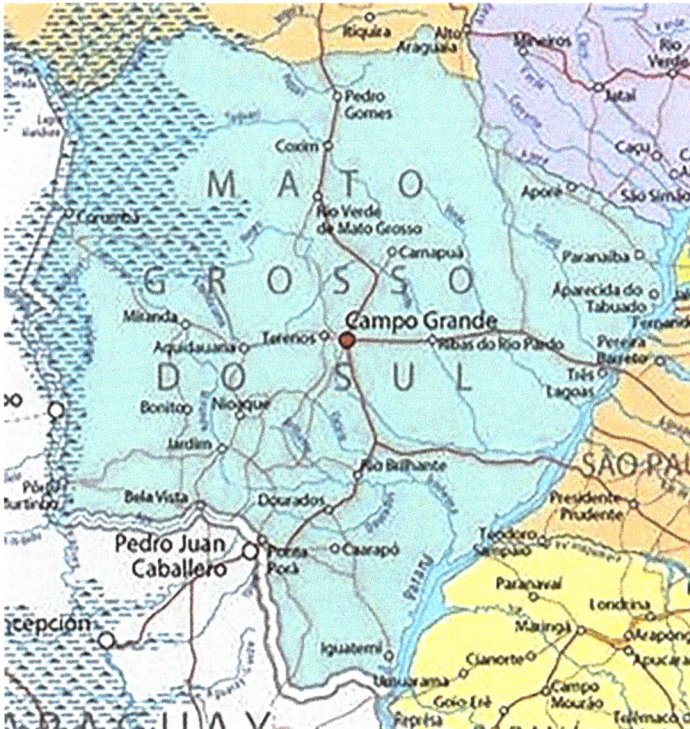

Coxim is located about 300 miles north of Campo Grande. Source: Free map from Maphill. 2011

1_____200 miles

In 1966 Coxim was on the only north-south road between Campo Grande and Cuiaba. It circuited the massive Pantanal swamp that occupied most of the western part of the State. It was a stopover for people wishing to travel to either place. I had to wait until eight that evening for the next bus.

I proceeded to entertain myself by visiting various friends I knew in town. Over the past months I had made many contacts in Coxim and among them were some young educated southern Brazilians who had come from Sao Paulo to this outpost to staff the local branch of the Bank of Brazil. They were always looking for interesting people to visit with and I must have been one of those.

Elhio, one of my banker friends offered his small apartment as a place for me to relax and take a rest. At two PM, during the heat of the tropical day I decided to take him up on his offer and took a nap on his bed. Elhio's apartment was functional; a shared bathroom, down the hall, clothes strewn around and I could tell it had been a while since the laundry maid had been around.

About hour later I woke up scratching. Something in his bed liked how I tasted! Itching all over I finally realized that Elhio's fleas had just finished a meal. I decided that this was not the best place for me to be resting.

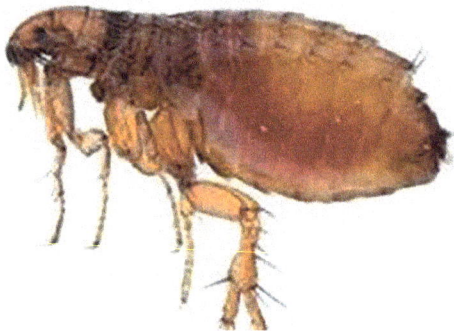

Common flea. Fleus Home, fleas, /cdc.gov

As the late afternoon began to cool off, people became active in the street and I joined a group playing basketball at the town park for a few hours.

One of the basketball players invited me to join his family for a dinner of rice and beans and roast beef. After dinner I went to wait for the bus. I stowed my bag in the overhead and settled into a seat and went to sleep, knowing that this would be an all night trip and I would arrive in Cuiaba in the early morning.

About an hour into to trip, I awoke to find myself itching all over again. Whatever I picked up from my friend's bed had stayed on my body and now they were taking their next meal.

Fortunately, I had my Peace Corps medical kit with me which had some flea powder that I dug out and shook powder around in my body and in my clothes. It worked! the flees stopped biting!

Peace Corps medical Kit

As I was relieved from the fleas, it seems that the fleas found an alternate host! The guy next to me began scratching. I was about to offer him my flea powder when he decided that I was not a good one to be sitting next to and got up and moved to another seat, taking the fleas with him!

8. GATOR FOR THE GATORS

While living on the University of Florida Campus, as a graduate student during the early 1970s alligators were around but did not present much of a threat to people. As opposed to crocodiles, it seems that they do not like human flesh!

Four-foot alligator along Lake Alice. Alligator geniest die Sonne by Sebastian Fuss is licensed under CC By-SA2.0

Alligators live in Lake Alice on the south side of Campus and like in the Everglades, they are fish farmers. They were no problem until their lake was invaded by the invasive water hyacinths that covered the water and sucked out all the oxygen. This did not directly bother the Alligators, but it did starve the fish of oxygen which in turn starved the aligators of their traditional food.

The next thing we knew, "Gator Country" seemed to have expanded its magnificent logo and gators were wandering campus, looking for alternative food. Fortunately, that was not human beings

Lake Alice covered with Water Hyacinths. Photo by Rick Bein 1973

I recall an alligator encounter on campus when a woman was walking her small dog past a wooded area. Suddenly she looked back and discovered she had an alligator on the leash instead of a dog! Needless to say, she let the alligator keep the leash. Not long after that the University began dredging the water Hyacinths off of the lake restocked it with fish. Consequently, the alligators reduced their nightly adventures.

Fortunately for people, alligators find dog meat rather tasty! Fences around back yards are not there to keep the dogs in, but the gators out!

The alligator (Alligator mississippiensis) is a keystone species in Florida. It has a wide ecological niche which many other species depend upon for their survival. American alligators are found in the southeastern United States including all of Florida, Louisiana, the southern parts of Georgia, Alabama, and Mississippi, the coast of the Carolinas, Virginia, Southeastern Texas, southeast corner of Oklahoma, and the the southern tip of Arkansas. Not to be confused with crocodiles which rarely appear in the United States, the Alligator does not live outside of the USA.

Freshwater is its preferred habitat, while it does venture on to land when necessary to find food. American alligators live in freshwater environments, such as ponds, marshes, wetlands, rivers, lakes, swamps, as well as in brackish environments. The Florida everglades provides the most ideal habitat and that is where most "gators" are found.

Florida Everglades map by the US National Park Service, Free maps, npmaps. com. Notice the small green linear hummocks shown in the central north section of the map.

The Everglades is a gigantic marsh of mostly shallow water extending for miles across south Florida, and, if it were not for the alligator, it would be perfectly flat with only three feet of standing water. Alligators create small islands in the everglades which are called hummocks where they lay their eggs. They make hummocks by excavating earth from under the water, deepening a spot to as much as 6 feet, while piling up hummocks one or two feet above the water. These small islands and pools actually serve a wide habitat for small mammals, other reptiles, birds, and insects as well as fish. The alligator is therefore considered an important species for maintaining ecological diversity in wetlands.

The hollowed out deepened pool is ideal for fish to breed and live comfortably. Essentially, it serves as a fish farm for the alligator who is the manager of this small ecosystem, harvesting fish from time to time but always allowing a breeding population to survive. This symbiotic community thrives sustainably.

Not many universities have the namesake of their logo actively living on campus. (University of Florida is locate in Gainesville in North central part to the State). When athletic opponents come to challenge the "Gators", little do they know with whom they are really competing?

9. BOT FLY

Deal with Health Issues where They Arise.

I always say, when you are in a place and a health issue comes up deal with it at that place. It is there that the locals understand its nature, what causes it, and how to cure it. Case in point occurred when Mary, my first wife, and I were in Campo Grande, Mato Grosso, Brazil where I was conducting my dissertation research.

A bot fly had laid an egg in Mary's thigh and it matured into a grub with a little breathing hole that it kept open. Normally these parasites choose spots on the backs of cattle where it is exposed to the sun. After about six weeks the grub matures and metamorphoses into a fly, which crawls out, allowing the wound to heal. The hole heals after that.

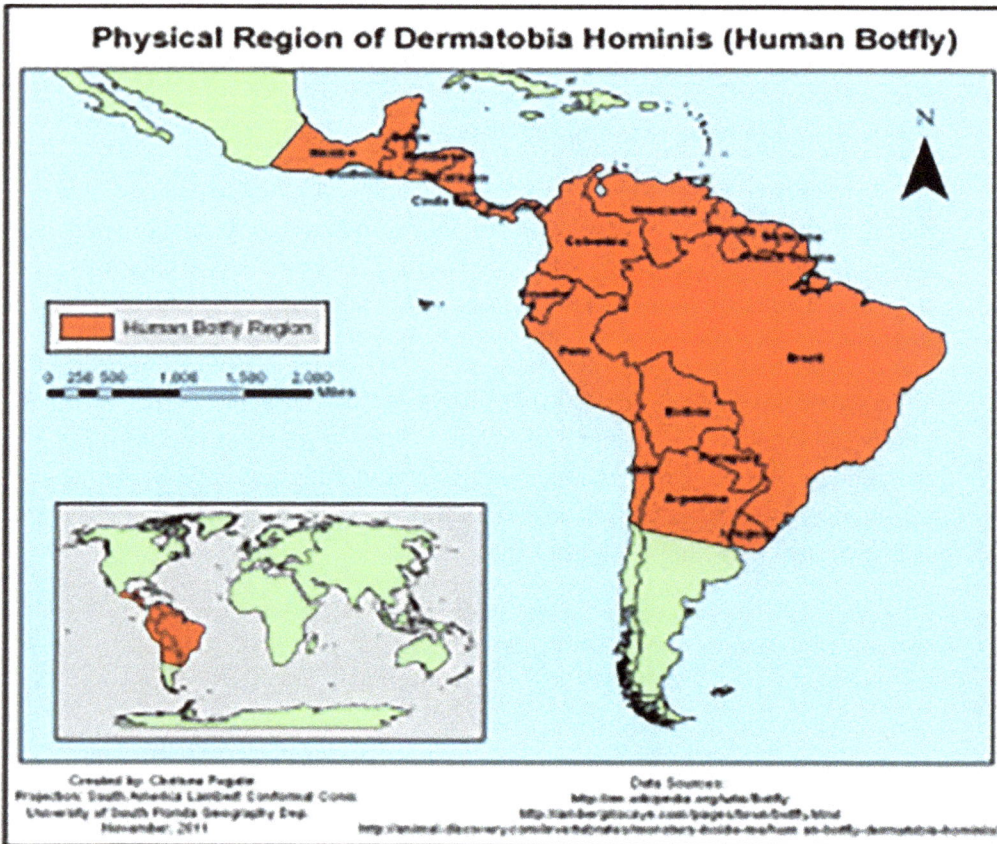

Physical Region of Dermatobia Hominis (Human Botfly)

We did not know what the little hole in Mary's thigh was, and we thought it would heal soon. It looked like an insect bite; it was swollen, oozed but refused to heal regardless of the medication applied to it. The sore continued into the second week and was not any better.

Larva of the bot fly, Dermatobia hominis (Linnaeus Jr.), lateral view. Photograph by Lyle J. Buss, University of Florida 2008.

That weekend we were invited to a friend's cattle ranch and since Mary was wearing shorts, the lady of the house recognized the sore as a cattle grub larva. We asked, "How do you get rid of it?"

Her husband the rancher had the prefect answer. He got out a slab of bacon and taped it tightly over the hole. In a matter of minutes the grub was feeling suffocated and proceeded to extend its hole up through the inch and a half of bacon. With in an hour the grub had moved out of her leg and into the bacon. The rancher simply removed the bacon with the grub in it and threw it away. Within two days the hole had closed and a week later it was completely healed.

Best Slab of Bacon Ever by Tom Ipri is free licensed under CC BY-NC 2.0

In contrast to this simple home remedy, there was a USAID couple from Texas working in Campo Grande and the wife, a registered, nurse, got a cattle grub on her chest just at the top of her cleavage, another place that received a lot of sun light. When Mary told her what it was, she was extremely grossed-out and went into a great panic. No home remedies for her! She wanted the best that medicine could provide and in her opinion that was not in Brazil.

She caught the next plane back home to Houston, where she knew the doctors could fix anything. At the University of Houston, the physicians had never seen such a thing and at first were not sure what it was. But they were fascinated with it and wanted to study this strange little alien just under her skin. After about a week they finally figured it out and staged a surgery for the medical students to observe. With an audience of over thirty people staring at her chest, they cut her open and extricated the greasy grub!

It took her a month to convalesce when she eventually got the stitches out. She refused to go back to Brazil after that.

Now, which of these two curing solutions produced the least trauma, the least expense and the least negative attitudinal memory?

10. WHO IS DOING THE DISHES?

Who Is Doing The Dishes?

In the 70's I was working on my doctoral studies about land tenure in southern Mato Grosso, Brazil. I was traveling from one small farm to the next as I completed my survey in the Terenos Agricultural Colony. Many of the colonists had sold their depleted lands which were then combined with other small farms to form cattle ranches. I was studying about the land use practices of the Terenos farmers and how they managed their resources. The remaining colonists were all poor, their average holding not exceeding 15 hectares, and most were looking for new forest land to clear.

I walked onto the property of an old native Brazilian, and after 100 meters came to the dilapidated structure of a house. I clapped my hands, as was the custom of getting some ones attention in much of rural Brazil. From the far side of the house he called out for me to come join him. A pack of hounds echoed a few barks then settled down.

He was sitting beside a campfire where two pots were warming. He had been whittling on a piece of wood, but when I arrived he was pleased to put that down and spent some time visiting with me. He was happy to have a visitor and he pulled a bucket of water out of a nearby well and started a pot of coffee. We chatted on as I worked my way through the questionnaire.

Unlike the other agricultural colonists who had cleared their land, he had not done much with his 15 hectares of woods. His small patch of forest served as source of fuel and an occasional place to hunt. He helped his neighbors from time to time who paid him with a sack of rice or beans.

The colonist poured me some coffee in a beat up old cup. He told me that he enjoyed hunting with his dogs, but his own property was too small, so he frequented some of the larger areas of remaining forest. I had noticed that the six hunting dogs lying listlessly in the shade were not well cared for as they were underfed and had numerous sores on different parts of their bodies. Many of the sores were infected and gave off a bad odor. I asked him why his dogs were so thin. He told me that hungry dogs made better hunters.

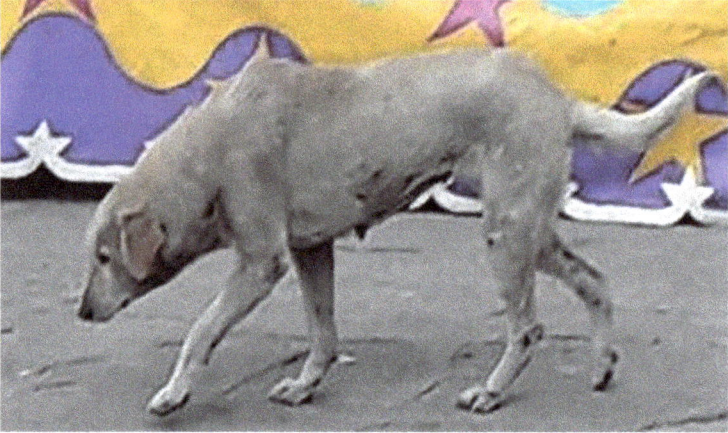

Example of a hungry dog. Photo by Rick Bein 2018

As it was late morning and he had answered all my questions, I was preparing to leave. He insisted that I stay for lunch since it was already warm on the fire. It would have been rude to refuse and since I did not have any other plans, I agreed to stay. He apologized that the food was not much, but he wanted to share with me. At that moment the sleeping dogs smelled food and suddenly came alive and began milling around, snapping and snarling at each other. Their stench was revolting and I was starting to lose my appetite.

"Take a plate off that shelf!" as he pointed to a stack of metal plates sitting at the bottom of someone's discarded bookcase. I chose a plate and a spoon. As I suspected, the pots on the fire contained typical Brazilian fare, rice and beans.

I dished up my plate and sat down on a log. He scolded the dogs who were competing to get close to us and they quieted for a moment. As I began to finish my meal the dogs again became agitated.

The old native noticed that my food had been eaten and offered me a second helping. When I declined, he took my metal plate and tossed in the ground. There upon, all six dogs descended on the plate, snapping and clawing at each other for the privilege of licking it clean. The plate was scooted back and forth around the small yard and soon no morsel was left. His plate also went on to the ground whereupon the dogs gave us a repeat performance.

After the hubbub was over and what might be considered their meal, the dogs retreated to their earlier territories in the shade and went back to sleep. The licked plates remained where the dogs had left them. Thinking I could be of help, I picked them up and asked "Where can I put these dirty dishes?"

"No need to do that; they're clean, just put them on the shelf where they came from!"

If there was one consolation, it was better to know after I had eaten, who the dishwashers were!

11. MARRARA

THE WORST THING I EVER ATE

By

Rick Bein

It was 1974. I had been in Khartoum, Sudan for a couple of weeks when I had the opportunity to travel into the rural area. Chris Winters, another American Geographer teaching at the University of Khartoum for the past year in Sudan, arranged for the trip to a village a two hours out in the desert. The villagers were honored to have American professors in their homes. Our host family prepared to have the best food and accommodations possible for us.

Sudan-administrative-map Copyright © 1998-2021 nationsonline.org. This village sits 30 miles north of Khartoum along the Nile.

Waiting for dinner, our host and several of his friends kept us company while we tried to be the best of guests. A plate of "marara" was served while the main food dishes were still being prepared. Marara is a traditional hors d'oeuvre in parts of Sudan and is made up of raw sheep organs covered with chili powder.

Example of Sudanese Marara made of raw sheep entrails tempered with slices of onions. Photo by Abdulaziz Muhamad.

Our host gave me first choice of this variety of delicacies. I looked with alarm for Chris, but he somehow had disappeared, and I realized that I was to experience this new situation alone without guidance.

My highest choice would have been to thank our hosts and not take anything, but I felt that I would offend them by not being a gracious guest. I rationalized that "Maybe the chili powder will kill any living parasites or bacteria."

I quickly scanned the array of fresh body temperature entrails including liver, heart, kidney, testicles, gallbladder and an unknown that I carefully chose because it seemed to be the smallest and least disgusting. (I learned later that is was the epiglottis) Not wanting to delay the process by thinking about it too much and becoming sick with the thought of it, I popped the inch long organ into my mouth and began to chew. I wanted to get this over with and if I could break it down to a size that I could swallow, I would be done with this little ordeal.

To my dismay I found this tiny organ was like chewing on an inch of garden hose and what's more when it came in contact with my saliva it began to swell. Finally, I bit down harder than usual creating enough pressure to squeeze out some kind of bile which squirted into my throat.

Desperately, fighting the urge to retch, as my mouth became fuller and fuller, I frantically tried to think of a way out. Finding none which I thought would keep me in the good graces of my hosts, I took one desperate gulp and the mass of glop slowly slid down my throat.

I had passed the test. However, my hosts were delighted to see how quickly I had taken to one of their delicacies and eagerly offered me another piece. There was no way I was going to repeat this performance and I was highly relieved when my host accepted my sheepish "No Thank you".

Back in Khartoum I related the experience to some of my Sudanese friends. They proceeded to gag and laugh and explained that they never eat marara and informed me that in fact, it would not have been considered offensive had I refused to eat it in the first place!

12. CAMELS MILK

Camel's Milk is not my highest choice.

While I was on the faculty at the University of Khartoum one of my Sudanese colleagues invited me to go on an excursion northeast of Khartoum and east of the Nile River.

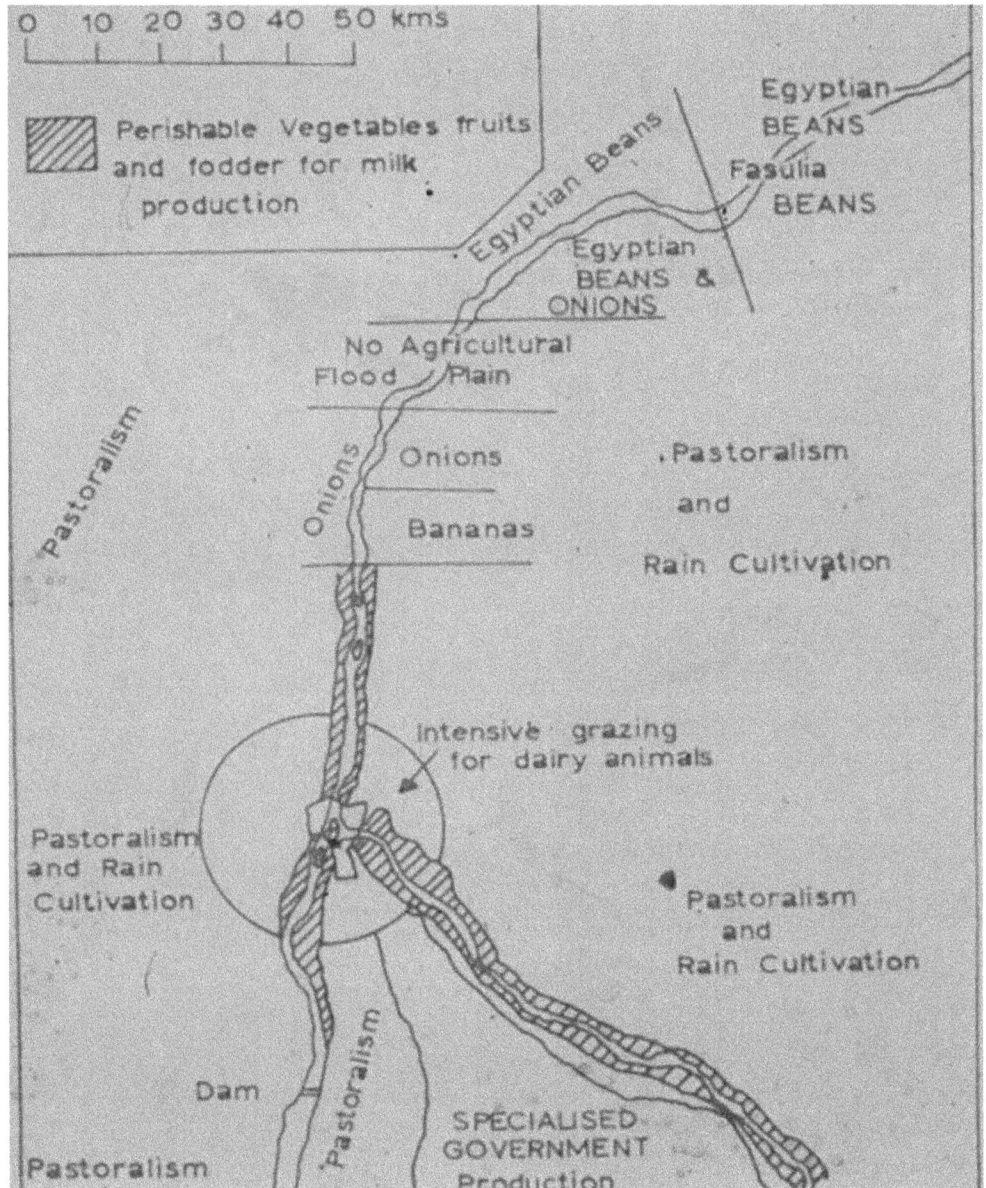

0 10 20 30 40 50 kms

Perishable Vegetables fruits and fodder for milk production

Egyptian Beans

Egyptian BEANS

Fasulia BEANS

Egyptian BEANS & ONIONS

No Agricultural Flood Plain

Onions

Onions

Bananas

Pastoralism

Pastoralism and Rain Cultivation

Intensive grazing for dairy animals

Pastoralism and Rain Cultivation

Pastoralism and Rain Cultivation

Dam

Pastoralism

Pastoralism

SPECIALISED GOVERNMENT Production

The Butana desert is located in the areas designated as Pastoralism and Rain Cultivation east of the Nile.

Map scale: 0_____100_____200_____300 miles

Camels grazing thorn bushes 1974. Photo by Rick Bein

Suddenly a man came running out of the bush toward us waving his hands. We waited for him, and he came up to the driver's window and said: Do you have any water?

Sure, we do. Do you want a drink?

Yes, please! He responded.

We got out of the vehicle and opened the back where we had five-gallon container of water. We poured a glassful and the camel herder drank it as if he were in heaven. After a few more glasses, I asked. How long has it had been since you had any water?

"Six weeks".

"How have you survived all that time?" I asked

"Camel's milk" he said as he pulled a little bag of liquid hanging from his belt.

The favorite camel. Photo by Rick Bein 1974

It seemed he had not talked to any humans for six weeks also and to entertain us got his milking camel for us to see. I curiously asked if I could try some of the milk and he said "sure" and handed me the leather bag. I smelled something a little rancid, but I had to try it. I took a tiny sip, and it was enough! It tasted awful, like spoiled fermented milk. I quickly handed it back to him.

The look on my face must have humored my Sudanese colleagues as they all they had a good laugh. They knew what I was gagging about. To make matters worse, that tiny sip had repercussions that lasted several days.

Interesting enough, we have learned one strategy for humans to quench their thirst in the desert, but how do the camels do the same? Camels can go many weeks without water, but in the Butana Desert the camels found that they could also survive on the water contained in the tiny succulent fruits that they harvested among the thorns of the desert bushes.

13. FEASTING WITH THE NYANGARA IN SOUTHERN SUDAN

About 50 kilometers north of the Uganda border a small crowd of Yanguara tribesmen from southern Sudan hungrily watch, politely waiting for me to do something. A large bowl brimming with roasted flying termites (Macroterms spp.) sits before me. I pride myself with being able to eat almost anything that someone else can eat, but insects? They look like dark reddish brown half inch sticks with wings sticking out.

I am thinking: "This is not the going to be easy!"

Termite preparation by Mercy Uchenna YouTube 2008

South Sudan regions map.png by Peter Fitzgerald is licensed under CC BY 2.5

"Patrick, what am I supposed to do?" Patrick Ladu, one of my students from the University of Khartoum is very patient with me. His tribesmen are also patient.

"Take a few off the top of the pile and put them in your mouth! You are the guest of honor and you must eat first. No one will eat until you do! This is a delicacy as they have prepared the very best for you!"

Mustering my courage, I pick a couple of large two-centimeter-long bugs off the top and before I can change my mind, put them in my mouth. Immediately the crowd dives into the community bowl of termites, filling their mouths with ecstasy.

I am overcome with their delight at what they are eating, forgetting what is in my mouth, and I begin chewing. To my amazement, I find the termites pleasingly salty with a pretzel like texture. I try to grab a few more, but that is my last bite as the pile disappears in the frenzy!

This is an East African delicacy that is craved for its saltiness. The heavy tropical rains of Southern Sudan, Uganda and Western Kenya wash most of the natural salt out of the soil. As a result the salt content in plants is also low. Termites obtain minerals found deep in the subsoil where they dig their burrows and salt being one of them is concentrated in their bodies, making them a high demand item by the local people as they seek to meet the salt deficiency in their diet.

In the mid-1970s they were a great snack food. People walked around the East African markets with paper cone wraps full of roasted termites munching as they went along. Instead of eating popcorn at the movies, they snacked on termites!

When I returned to Kenya in 2007, I noticed very few people eating termites and those that kept the habit were in small villages.

African termites build huge hard clay mounds up to ten feet high in which they live. They process decaying vegetative material in the soil, eating the plant material and depositing the mud-like material to form the mound. In this area of Africa, termites are welcome and when a new termite mound starts to appear, the first person to see it puts a claim on it along with the right to collect them.

Termite mound with author, Photo by Rick Bein 2012

Some of the termites swarm in the evening following a rainstorm and fly off to start a new colony. The local people know when this is going to happen and just before dark, dig a shallow rectangular hole measuring fourteen by fourteen inches square and about two inches deep at the base of the mound. In the middle of this small pit a candle is left burning and when the termites are attracted to the light, they burn their wings and fall into the shallow pit. In the morning the pit is full of crawling termites which the owner collects.

Over the next few days, I have many chances to eat termites. Besides roasting them, they are fried, baked, broiled, boiled or made into meatballs. A recipe book could be written!

I Actually started to become accustomed to the salty flavor of the termites. The one thing I did not get used to, was the wings and antennas catching between my teeth that poked out when I smiled!

I found that the Nyanguara used their shifting cultivation fields as bate for wild animals. They set snares at the entrance to the fields. One day a Thomson Gazelle (Gazella thomsoni) was caught and although it was a rather small critter, it was easily made into a meal for a couple of families.

Gazelles grazing, Photo by Rick Bein 1976

A one-ton cape buffalo (Sycerus caffer) was caught a few days later. The huge carcass was shared far and wide throughout the rural area. Even then the people could not eat it fast enough. With no refrigeration, the meat began to spoil. Being conservative with their resources, the people did not want to waste it and kept re-cooked it even as it turned rancid. Yet, the buffalo meat sat in the stew pot over the cooking fire. When the fire went out at night, the meat continued to decompose

Watering place for the Buffalo. Photo by Rick Bein 1976

At first, I enjoyed eating buffalo, but after a few days it began to taste bad. I was not sure who was going to eat it first, me or the maggots.

Finally, I said. "Patrick, I cannot eat any more of this awful buffalo, I think we should leave and go on to visit some more of your relatives."

Patrick agreed and the next day we set out walking. Ten kilometers later, it was mid-day and we stopped at a village just in time for lunch. Lunch, you guessed it, was another shared portion of that same rotting buffalo!

14. LIONESS OF MASAI MARA

It was June-early July, 1976 and I was on vacation in Kenya from the University of Khartoum in Sudan. I was traveling alone as Mary; my wife had chosen to take the twins and go back to the States for our biannual home leave. The university had given us just enough money for two adults to fly. We found that the two three-year olds could fly for the price of one adult.

Natural Vegetation in Africa

Traveled from Sudan, southern Sudan to Kenya on African Vegetation Map.
Perry Castaneda 1986 Public Domain map.

I chose to stay in Africa and complete some field work. After a tearful goodbye at the Khartoum airport, we got on separate planes; the family going to the States and me to Juba in South Sudan.

I spent two weeks in Equatoria Province in Southern Sudan (It was part of Sudan then) where I met with my southern Sudanese students. They escorted me to their villages where we interviewed farmers.

When the field work was completed, I traveled by group taxi to Nimule on the border with Uganda. I had intended to travel by land to Kenya via Uganda, but after crossing into Uganda and traveling a few miles I heard some frightening news.

A man on the bus informed me, "There is a failed coup attempt and Idi Amin is blaming America and the CIA. He is screaming about Americans being evil and wants to rid Uganda of them." I confirmed this with the driver and several other passengers.

Hearing that, I realized that I would probably be detained when the Ugandan police saw my passport at the next check point. I immediately stopped the bus and got off and hitch hiked back twenty miles back to Sudan. Any American was at risk and I did not want to be one of them.

Once I arrived in Juba, I inquired at the airport about flights to Nairobi, Kenya. Ticket salesman, said "Yes there is a plane in the morning to Nairobi, but it stops in Entebbe, Uganda Airport for a couple of hours."

"When is there a direct flight to Nairobi?" I asked.

"There are none, all flights have stop over's in Entebbe"

"Would I be a risk in the airport?" I asked.

"You would be much safer there than traveling by land. You will only be there a short time and since you will be in transit and won't be leaving the airport."

I did want to get to Kenya and I decided to risk it. The flight to Entebbe was uneventful and the airport was surprisingly immaculate and clean. It looked new.

"Yes, The Israelis finished building this airport last year" I heard someone say.

The plane to Nairobi was full of multicultural passengers, dozens of different nationalities who also found themselves transiting through Entebbe. In the seat next to me were some Italians and we began trying to communicate. I knew no Italian and they spoke very little English but we found that we could get by in Spanish.

As the Italians and I got off the plane and went through customs, we started talking about things to do together. They agreed that I could join them. As it turned out they had done very little planning with regard to an itinerary and so I told them what I had in mind and they agreed to join me in my adventure.

The first thing we did was go to the "message tree" in Nairobi, famous at that time among people who were backpacking, traveling on low budgets or who were "in" the hippy crowd. People left notes on the trunk of the tree to connect travelers with similar interests. We found nothing of interest in the tree but took in the ambiance at this collecting point and enjoyed a few Tusker beers. Tusker lager is a rather strong but tasty beer, and it did not take long for me to become sleepy. Someone suggested a low-cost hotel and we went there to sleep.

The next day we rented a car (of course it had to be a Fiat!) and headed out to the famous Masai Mara National Park. It was a lovely day in July, and we headed south down the rift valley. A Masai man with his bright red blanket waved us down and indicated that he wanted a ride. The Italians were a bit leery about that, but I insisted, and I had him sit in the middle of the back seat. I was really curious about him and tried to make conversation, but he was not interested and just grunted. He also objected when I tried to take his photograph. Finally, one of the Italians asked me to leave him alone. After about 30 miles he indicated that he wanted to get out.

Kenya 1995 Lonely Planet Map

The roads were good, and we arrived at Masai Mara Park by mid-morning. We paid our money and drove into the park. Thousands of wildebeests (Gnus) were stampeding across a vast dusty plain of over-grazed grasses. This was their annual July migration. As they ran they seemed to do more leaping than they did running. In two bounds one of them could completely cross the two-lane road. Alongside the thunderous roar of thousands of hoofs, I could hear a persistent sound of frogs croaking. I finally realized that was not frogs at all, but the sound wildebeests make. It sounded like "croak, croak, and croak." I looked it up in the guidebook where it talked about the wildebeest saying its own name and so the local name became "gnu". I tried unsuccessfully to imagine how "croak" could be construed as "gnu."

Migrating wildebeests (Gnus) stampede on the semi-annual migration.
Photo by Rick Bein 1976

As the rushing hoard of thousands of gnus completely covered the landscape it seemed a little dangerous to move through the middle of them. We were afraid to drive across their path, but we saw other vehicles doing it and we found that the gnus avoided us and ran around the car. We proceeded for several miles a midst the rampaging animals; a very spectacular experience.

Twice a year the gnus follow the sun across the equator.

The wildebeest ran around our car and continued on their journey Photo by Rick Bein 1976

We found that the gravel road inside the park was not kind to the small Fiat tires and one of them had a puncture. We jacked up the car and had the tire off and were ready to put on the spare, when I had an eerie feeling and I glanced back into the scrubby bush beside us.

My eyes connected with those of a lioness! She had been watching from about 50 feet away. The eye contact seemed to startle her and she started loping toward us! With a quick word of warning, the four of us managed to get into the car in record speed. Amazingly, the frantic jarring of the car did not cause it to fall off the jack!

Once inside the car we checked to see if any of us had been eaten as we looked around to see where the lioness had gone. She had not come all the way to the car. We spotted her 30 feet back into the bush where she sat watching us, but then we noticed five lion cubs playing around her. She must have thought when I made eye contact with her, I had also seen the cubs and her intent was to protect the cubs. She and the cubs stayed put for another hour. So did we!

Author unknown. Photo available on www.google.com

Then the lioness and the cubs began to move parallel to the road. After about a half hour they walked onto the road ahead of us and continued for 300 feet before disappearing into the bush on the other side of the road.

Finally, we got up the courage to get out and finish changing the tire.

The do-nut spare lasted about half an hour before it also went flat. We had no choice but to wait alongside the road until help came. A truck eventually came and took me and the original flat tire several miles to a place where it was repaired. An English family in a four-wheel drive land rover drove me back to the Fiat and the waiting Italians. It was late at night before we made it back to Nairobi. They thanked me for the adventure but declined to go on any more of my excursions.

Interesting enough, as I spent a few more days in Kenya, news came that an Air France airliner had been hijacked by some Arabs and forced it to land in Entebbe airport on the 27th of June. That was just a day or two after I had left that airport. Because the hijackers had detained the Jewish passengers while releasing the other passengers, an Israeli response resulted. Operation Entebbe was a counter-terrorist hostage-rescue mission carried out by commandos of the Israel Defense Forces (IDF) at Entebbe Airport in Uganda on 4 July 1976.

It was a relief to be free of Uganda at that point and I decided to find another way to return to Sudan. I found there was a flight to Addis Ababa, Ethiopia, from which I could transfer to another flight back to Khartoum.

One of the more adventurous Italians decided to go with me to Khartoum. He thought he could arrange a connecting flight to Italy from there. Once in Addis Ababa we decided to spend a day exploring a bit. The next day we arrived in Khartoum, to find that his paperwork was not up to-snuff with the immigration officers. My passport and documentation showed that I was a legal resident and there was no problem, but some reason they became suspicious and arrested him and took him to jail. I felt somewhat responsible for not foreseeing any problem for him coming to Khartoum. I spoke to the police that he was just traveling in transit, but they told me to go home and they had things under control. Later I went to the police station to inquire about him and I got the same result. I never did get an answer and to this day, I still don't know what happened to him.

15. DOG BRAIN

Not My Highest Choice for Lunch.

When it came time to return to the States and leave Papua New Guinea in July 1999 Maryellen and I booked our flight on Qantas Airlines. We purchased the special round the world ticket that allowed six stops where passengers could disembark and stay for a number of days or months before re-boarding to continue our journey back to Australia. Our first stop was Manila, in the Philippines. After exploring the Island of Luzon, it was suggested that we take a two-hour boat trip to the Island of Boracay where tourists normally visited.

Snobbishly we ignored the activities on the beautiful beach and began looking for something more typically Filipino in culture. We spotted on a map of Boracay, a museum located on a hill overlooking the ocean, and decided to check it out.

Location of Boracay. Maps online free www.pinsdaddy.com

It was a bit removed from the regular tourist area and as there were no cars on the Island, we hired a motorbike taxi to take us there. Motorbikes were quite common all over the Philippines and Boracay was no exception. We told the bike operator where we wanted to go and climbed on when he took off. As it turned out the weight of three of us was a bit much for the bike on the steeper parts of the hill, so we had to walk.

He came to a stop in some remote place and pointed toward a trail leading into the forest. After fifty yards we came to a building on stilts. Climbing the wooden steps up to the veranda we found the front door was locked and the place seemed to be deserted. We walked around the building and found two curators enjoying their lunch. They were so intent on their food that they hardly noticed us approach. They were eating out of the same container with their chopsticks and as we drew closer, we saw that they were eating what looked like a brain out of a sawed-off skull!

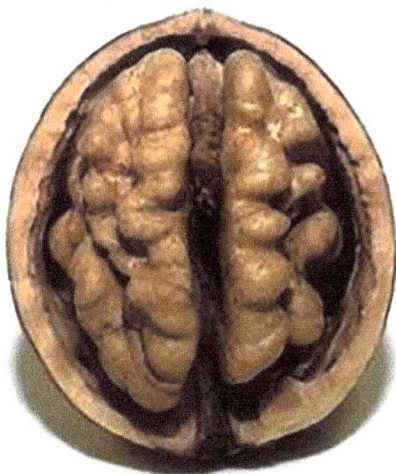

Cracked Wall nut simulates how the brain looked.

They were patient with our intrusion and offered us something to eat. Maryellen immediately did a "180" and headed back around the building! My curiosity was too much, and I asked what they were eating.

"Woof-woof" was the answer. Even though she was trying to get out of earshot as quickly as possible, Maryellen still heard their answer and after seeing the expression on her face, I decided it would be a good idea to decline the offer!

When I caught up to her, "You just had to ask!" was her comment.

Once the delicacy was finished the curators came and opened the museum. We paid a token amount and spent the next hour looking at a variety of costumes, jewelry, and other artifacts stacked on benches along the wall.

16. ADVENTURES WITH SEA TURTLES

Sea Turtles nest on most tropical beaches around the world. Many countries are now engaged in preserving these endangered species. This effort is long term since many of the turtles do not sexually mature for decades. The Leatherback, for example, only begins nesting when it becomes 30 years of age.

My first experience with sea turtles happened when I was working in a coastal village in Papua New Guinea (PNG). I had always heard of sea turtles as an endangered species, but never from a firsthand perspective. I had been working on contract with the NGO Village Development Trust (VDT) whose purpose was to conserve environments in Papua New Guinea. They had established an introductory relationship with the village of Lababia located forty miles south of the town of Lae along the east coast on the Solomon Sea. Their approach was two-fold: 1) create a scientific research station and 2) develop an eco-tourist program.

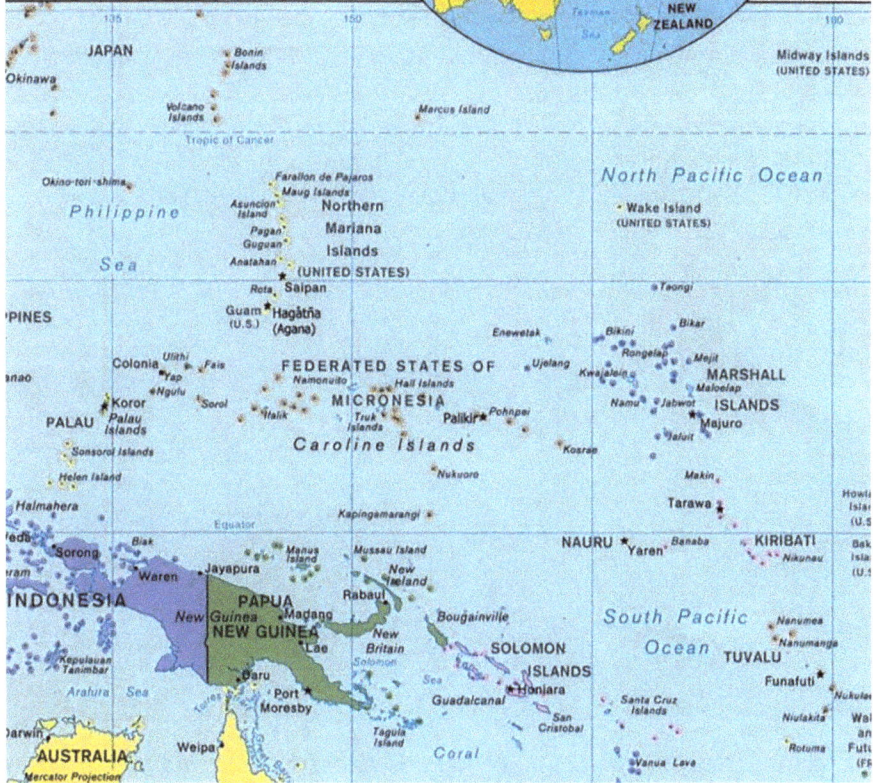

Papua New Guinea, by the Australian Geology survey. 2007.

As a visiting professor at Papua New Guinea University of Technology (UNITEC) in Lae, Papua New Guinea in Lae I was on an Asian Development exchange Grant to develop an Environmental Research Management Center on campus. My mission was to develop an environment focus among the academic departments through their teaching and research programs. VDT contracted me to build a research transect from the Lababia Village extending up the 6000-foot Blue Mountain behind. Being a pristine rainforest, the transect would provide a path for biologists to sample the unique fauna and flora.

Over the next two years my efforts extended beyond the mountain transect and I became involved with the Kamiali people, the residents of Lababia Village. The Kamiali thrive along their 15-mile-long coast engaging in fishing and gardening. The villagers invited me to experience much of their way of life. At one point, I discovered that their extended beach was a nesting habitat for Leather back turtles.

My visiting stepdaughter, Cindy Riggins and friend observing a nesting turtle on the Kamiali's beach. Photo by Rick Bein 1997.

I spent some time observing the female turtles coming on the beach and creating nests, laying their eggs, and returning to the ocean. I also became aware that none of the eggs ever hatched because the villagers dug them up and consumed all of them. I asked if any hatchling turtles were ever seen. The answer was "no."

Villager standing beside a nest, showing turtle eggs impaled on a stick. Photo by Rick Bein 1997

I decided to consult Levi Ambio, with whom I worked closely building the transect. Levi explained that the Kamiali considered the turtle eggs their gift from the universe and that it was their right to eat them. I asked him how the turtles would continue to nest if there were no new turtles to replace the old ones.

That notion struck home, and we began talking about how the villagers could give up this annual treat. He thought that would be difficult, but he said he would establish a no-harvest zone along a 100 meter stretch of beach right next to the village where he could keep watch. I was doubtful that this would work but worth a try.

I had to return to the States shortly after that and forgot about his endeavor. I went back to my teaching and research. I was awarded a travel grant to return to PNG the next summer of 2000, to follow up on my research on the food gardens of Kamiali.

When I arrived at the village, Levi greeted me with the news that for the first time in his life he witnessed hatching turtles marching down to the ocean. I was thrilled that I had made an impact.

After that, I had little contact with PNG, and continued with my work at IUPUI. A surprise handwritten letter from Levi arrived in 2007. In it, he explained what was happening to him. See copy of the letter.

[06 Sept 2007]

My Brother Rick-

It's a very long time since I heard from you & your family. How's you. I hope to see you some time. Also how's your wife and girls and Alex. Back here we are ok.

Nancy is married with two children- and Owen is also married with one child, I am also growing old, hope you are still young.

Brother, I have been trying to get you for years, where wee you from your last destination.

Rick, thank you so much for everything that you thought me. I am now a Leather Back Turtle Specialist here in the PNG. Enock is right now working with DR. Allan Asison as a field person. Tanny will now be a leader of the Blue Mountain track building from Tabare to the CliffSite Camp.

I have been attending International Sea Turtle Comference's as far as Honolulu- Australia- Solomon- and so fouth. I am ok now, using every knowlege you thought.

Brother, Dr. Allan is now building a Research Station here at Kamiali- and three other Branceh station along the trail to the blue Brauntains.

I am very Confident that DR. Allan will be very Sucessfull

My Brother, I still remember you and your family and all the Jokes.

iF this letter gets to you - you can reply using V.D.T. adress - they will always send to me.

Just a question - Have you thought of Coming back to PNG for a visit, if you're thinking of Coming back then write and tell me.

Bro, I think that is all for now.

Greeting from Levi family - and bye.

hope to hear from you soon.

L·A·

Letter from Levi Ambio 2007

My second experience with sea turtles was a major hands-on adventure. The ANAI Marine Turtle Conservation Project presented an opportunity to work with sea turtles in Costa Rica in 2001. They advertised for a university professor to lead a group of student volunteers to work in a turtle conservation habitat on Gandoka Beach, Costa Rica. Gandoka was located on the Caribbean side of Costa Rica, bordering Panama. I recruited a team of seven students who enthusiastic enrolled in the adventure.

CIA map of Costa Rica. The ANAI Marine Turtle Conservation Project located southeast of Puerto Limon on the Caribbean coast on the most southerly point next to Panama.

Costa Rica has taken the lead in protecting our terrestrial environment. Even though they are a small country they have invested a major effort in developing their country with a focus on the environmental conservation. Arriving at the airport, one faces about six recycling receptacles starting with paper, cardboard, plastic, glass, and metals. Locals are accustomed to recycling and are ready to remind others when they fail to do so.

Former Cost Rico president, Oscar Arias was the brainchild of this effort. In every way possible they have adopted conservation practices, beginning with recycling and habitat preservation. This effort has fostered an economy around Eco-tourism which brings people from all over the world to experience rainforests, savannas, wetlands, mountain forests, tundra, and even beach habitats. Nesting habitats extend to the rest of the world and many species migrate annually starting in Costa Rica.

Sea Turtles are an example as they come momentarily to lay their eggs and then spend the rest of the year at sea. They are trying to stop this age-old practice of eating the eggs by establishing sea turtle egg nurseries on several beaches. The habitats are high maintenance operations and require significant labor. Costa Rica has invited environmental enthusiasts from around the world to volunteer at these habitats.

Spend Two Weeks this Summer in Costa Rica

The IUPUI Geography Club is sponsoring a meeting, discussing a volunteering opportunity in Gandoca, Costa Rica working on the Marine Turtle Conservation Project.

When: Thursday, April 12th

Where: CA 203

What Time: Noon

Dr. Rick Bein of IUPUI's Geography Department will be there to share his experience volunteering in 2001.

For more info contact the Geography Club at geoclub@iupui.edu

Flier promoting the work-study course

We arrived in early July 2001, and they gave us a week to tour some of the traditional Costa Rica tourist sites like Monte Verde and Arenal Volcano. Then we went to work at the ANAI turtle nursery at Gandoka. Initially that involved learning how to patrol the beach, keeping our distance from the turtles coming up the beach, identifying where eggs have already been laid. We patrolled in pairs and took shifts during the night recording where turtles nested and we had to keep an eye out for poachers.

A controversial movement claiming "the harvest of some eggs for food is in keeping with preserving sea turtle eggs" has emerged in the province of Guanacaste Costa Rica at the Ostional National Wildlife Refuge. Photos by Desarrollo de Areas Conservacion.

The turtle egg laying season in Costa Rica is during the months of May, June, and July. It is said that a leatherback turtle can lay up to sixty eggs three times in a season. Few of the eggs hatch, not to mention all of the hazards a turtle goes through to reach adulthood at 30 years of age.

The connect the words approaches the beach cautiously, looking out for predators, and seeing none it continues through the breaking waves on to the head of the beach, and on to the furthest extent of the sand. She chooses a site, and stirs up a 15-foot perimeter in which she digs a hole as deep as her flipper can reach. At that stage she ignores any disturbance that might distract her.

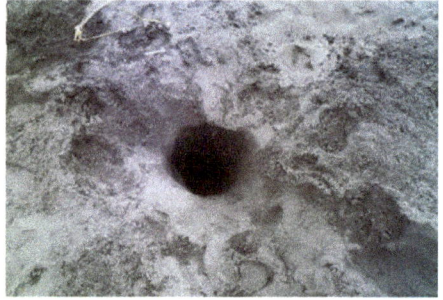

Leatherback leaving the water, Open nest exposed by poacher. Photos by Rick Bein 2001.

This nest was buried this deep, but a storm eroded the sand exposing the nest. Team member has to dig quite deep to reach the eggs. It is easier to place a plastic shopping bag in the hole to catch the eggs as the mother lays them. Photos by Rick Bein 2001.

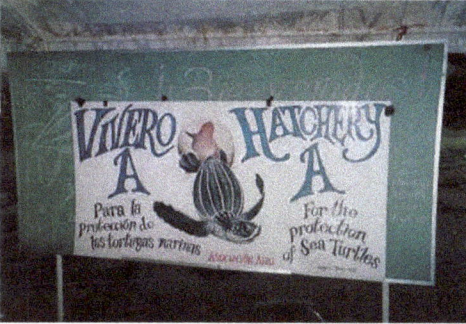

The eggs collected are moved to holes dug in the protected hatchery. Sandbags protect against high tides and storm surges. The enclosure prevents predators from reaching the eggs and plastic cages cover nests to contain hatchlings before releasing them to the sea.

A: A full hatchery midway through the season. B: Baby turtles coming out of the sand to be gathered and taken all together to be released. Photos by Rick Bein 2001.

Blue containers containing hawk-bill hatchlings being taken a few hundred

meters up the beach from the hatchery to avoid sea predators waiting near the hatchery for escapees. Hatchlings are released at the top of the beach to allow the march to the sea to strengthen their flippers for swimming. If released directly into the sea, they will drown. Pictures by Rick Bein 2001.

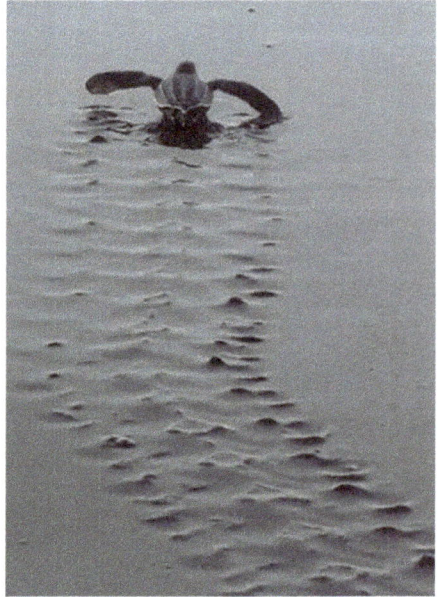

Hawk-bill hatchlings marching to the sea. Leatherback hatchling leaves a trail in the sand. Photos by Rick Bein 2001.

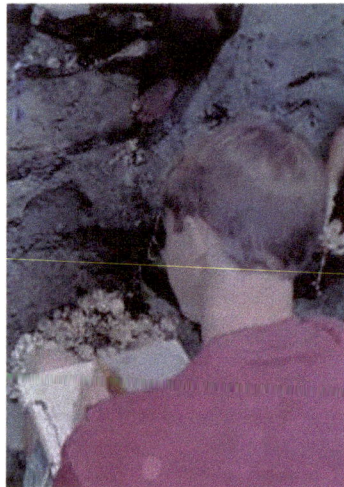

We assisted with the closure of the hatchery at the end of the season July 30, 2001. All unhatched nests had to be opened and eggs inspected. Records of the condition of the eggs were recorded. Photos by Molly Funk 2001.

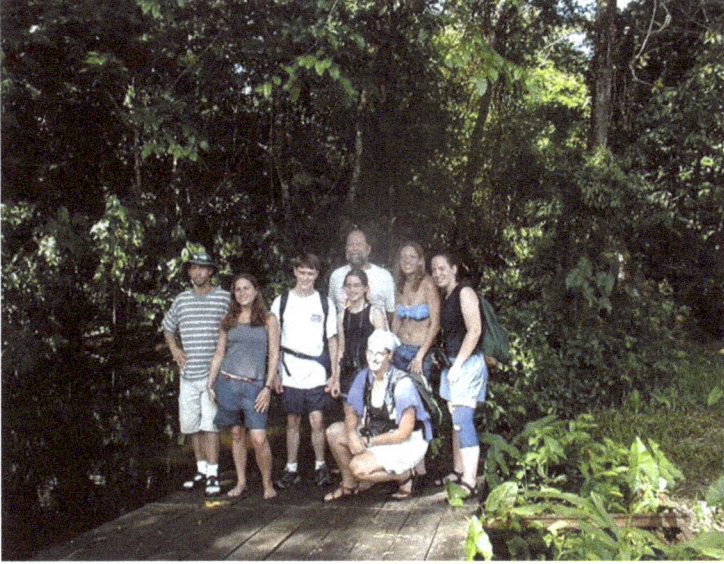

Indiana turtle conservation team, Costa Rica 2001

17. BABOONS OF CAPETOWN

The Baboons of Cape Town

Baboons around Cape Town, South Africa (at the southern tip of Africa) are semi protected. They are left in some of the upland areas away from the majority of people to fend for themselves. Most of these bands occupied secluded areas of Table Mountain and in the interior.

Map of South Africa with Cape Town highlighted (2011).svg by Htonl licensed under CC BY-SA 4.0 From Wikimedia Commons, the free media

Rick Bein at Cape Arugulas where two oceans meet. Photo by Jonathan, 2006.

They enjoy some of their native diet, but have become accustomed to eating what they can obtain from humans. Often the food is given but most of the time it is stolen. The baboons travel in packs, sleeping bunched together in the bush. They spend the day roving around seeking food. Garbage dumps are likely sites, and small animals like dogs and cats are targeted.

Open doors are favorites. A band of baboons can ransack a house in a matter of minutes, ripping down curtains, knocking over refrigerators and making off with large quantities of food. They usually leave deposits of excrement, some of which they smear on the walls and furniture.

The small, secluded, all-white town of Scarborough sits south of Cape town and Table Mountain along the Atlantic Coast about 5 miles north of Cape point. It is vulnerable to the ravages of the baboons which periodically raid the village. Doors are kept locked and people keep an eye out for each other with respect to the invaders.

The town of Scarborough hires a full time baboon watcher to shadow the local baboon clan and chase them away from homes and warn the people if the baboons still come into town. Even though this Xhosa man is physically fit to keep up with the baboons, he also has an old bicycle that he keeps close to where the baboons spend the night. Should the baboons trick him and get a head start into town, he is able to peddle on ahead to try and block their progress. Because they are protected by the government, he is not allowed to kill the baboons, but he can shoot them with a sling shot to discourage them from advancing any further. He keeps a pocket full of small stones as ammunition.

While I visited in Cape Town in 2003, I toured with my friend Jonathan, one of the locals, to Cape Point, a promontory that marks the south west point of Africa that has been made into a major tourist attraction. Magnificent cliffs tower over the roiling sea below. Thousands of birds nest in the crevices of the rocks not far from their source of fish that thrive in the up welling water. Typical tourist haunts with curios, souvenirs and snack food crowd the edges of the park.

After enjoying the sites, we retreated to the parking lot to get ready to return to Cape Town. Jonathan suddenly decided that it would be nice to have a snack for our return journey. He walked quickly back to one of the shops and bought some food.

He trotted gently back toward us, carrying a paper bag in his right hand. That was the moment for one of the local baboons who had been hiding in the bushes beside the parking lot. Running up from behind Jonathan, the baboon tried to snatch the bag from his hand. At the last moment Jonathan pulled the bag out of the grasp of the groping baboon and fled in between two cars.

Within seconds the baboon followed, leaping onto one of the cars and immediately onto to my friend's shoulders, straddling the back of his neck, holding on to his head with one hand and reaching for the food with the other.

It became apparent to Jonathan that this baboon was determined and was not going to give up without an even more serious fight. Besides, he did not like having the baboon's nasty rear end resting on the back of his neck. Tossing the food onto the ground was all that was needed to free himself as the baboon jumped after the food and began tearing up the paper bag. Fortunately for Jonathan, the only wounds were his lost pride at having been tricked by the baboon.

18. WILD ANIMALS, WHERE ARE THEY?

The Mozambique government wants to cash in on the kind of tourism that the South Africans have with Kruger National Park. Kruger is probably the foremost wild animal park in the world where you are guaranteed to see lions, elephants, buffalo etc. Limpopo Park abuts up against Kruger on the Mozambique side of the boundary where the South Africans have been donating hundreds of different surplus wild animals. But after three days in Limpopo Park, we saw only a few impalas, 3 kudus and a waterbuck. What happened to the other donated animals?

Waterbuck in Limpopo Park, Mozambique. Photo by Rick Bein 2005.

The 26 thousand tribal villagers who still live in Limpopo Park have not had much say about all these animals being released in their back yards, have gone to great extremes to protect themselves, their cattle, and their crops. After the lions ate a few cows and the elephants trashed a few fields, they were shot dead, and the rest of the wild animals have fled back into Kruger Park. Now the villagers have established who is at the top of the food chain, while acting in collusion with poachers. The government can't understand why the tourists aren't flocking in to spend the big bucks!

Farmstead in Southern Mozambique. Raised corn crib is a prime target for elephants. Photo by Rick Bein 2005.

Against this backdrop of four different stake holders: 1) the animal park managers, 2) Western donor agencies, 3) Mozambique government and 4) the local villagers, a much more complex situation is apparent. The objective of a National Park is primarily to provide area where nature can exist, protected from the over exploitation of human activity. It is the current view that there is an inherent advantage in maintaining natural areas to promote biodiversity as well as to support educational, recreational, and scientific research based on the continuous presence of wildlife species. The cost of providing natural areas may be provided through a number of scenarios. Governments frequently finance the maintenance of such parks, but under some circumstance's parks are maintained by their recreational users. I have seen plenty of instances in which recreational users are charged very high fees which are not put back into the parks!

A 2022 article from the Mozambique newspaper "Noticias" relates how new animals are continuing to be (re)introduced.

RESTOCKING OF BUFFALOES IN LIMPOPO NATIONAL PARKMaputo, 16 Sep (AIM) – A herd of 50 buffaloes has been imported into Mozambique's Limpopo National Park, as part of an ambitious restocking programme, reports Thursday's issue of the Maputo daily "Noticias".

About 3,000 animals of various species have been imported by the park, so far, mostly from South Africa. This is around half of the 6,000 targeted by the programme, that started in 2002.

Gilberto Vicente, of the park management, said that the buffaloes were brought into the country after going through all veterinary procedures to ensure that they are not suffering from any disease.

All the animals are being monitored, and "if there is evidence that any animal is affected by any disease it will be either slaughtered or put in quarantine", he said.

He said that the veterinary authorities will be monitoring the herds to ensure that they will not spread any diseases in the 25,000 hectares of the park.

Vicente explained that bringing in buffaloes is a pilot experience, to try and ensure ecological balance. Buffaloes play an important role in the food chain, he said, because they eat high grass, and open up areas for other herbivores to graze.

The Mozambican Conservation and Veterinary Committee, has expressed satisfaction with the strict observation of the restocking regulations, and hopes that this will be consolidated and extended to the so called "ecological corridors", that were opened with the removal of the border fence between Mozambique and South Africa.

The Limpopo National Park is part of the Greater Limpopo Trans-frontier Park, which also includes South Africa's Kruger Park, and the Gonorezhou park in Zimbabwe.

19. TRADITIONAL AGRICULTURE IN KENYA

F. L. (Rick) Bein

Department of Geography, IUPUI, USA

Gilbert M. Nduru
Department of Geography, Moi University, Kenya

Introduction on Kenyan food context

Agriculture is the main stay of Kenya's economy that employs over 75 percent of its labor force and is the lifeline for 85 percent of the population. It earns 60% of the countries foreign exchange. Agricultural production involves mainly mixed farming, the raising of crops and livestock. Cropping is more intensive in the high rainfall areas with maize being the main staple food crop and beans the most important legume. Coffee, tea, and sugarcane are major commercial crops. Seventy five percent of Kenyan agriculture is practiced by small holder farmers.

Climate

Kenya has a wide rage of climatic scenarios largely dictated by the highly diverse topography and moisture generating systems, hence creating a wide rage of agro-ecological zones. Two rainy seasons prevail in much of Kenya with the long rains normally occurring in March through May while the short rains come in September and October. Storms watering eastern Kenya are generated from the Indian Ocean while the Congo – Lake Victoria hydrologic system provides precipitation for western Kenya.

The high lands are well watered through orographic processes and the interior lowlands in the north are quite arid. The eastern and western high lands ranging from 1000 to 2500 meters in elevation receive abundant rainfall while experiencing cooler temperature with less evapo-transpiration. This results in varied agro-ecological zones that sustain a wide diversity of agriculture. See Table 1 and Figure 1.

ACZ	I	II	III	IV	V	VI	VII
Class	humid	Sub-humid	Semi-humid	Somewhat Arid	Semi-Arid	Arid	Very Arid
Rainfall in mm	1100-2700	1000-1600	800-1400	600-1100	450-900	300-450	150-300

Table 1: Kenya's Ago-Ecological Zones. Divided into seven agro-climatic zones (ACZ) based on suitable area for growing major food and cash crops. Source: Kenya Soil Survey 1982.

AGRO-CLIMATIC ZONE MAP OF KENYA

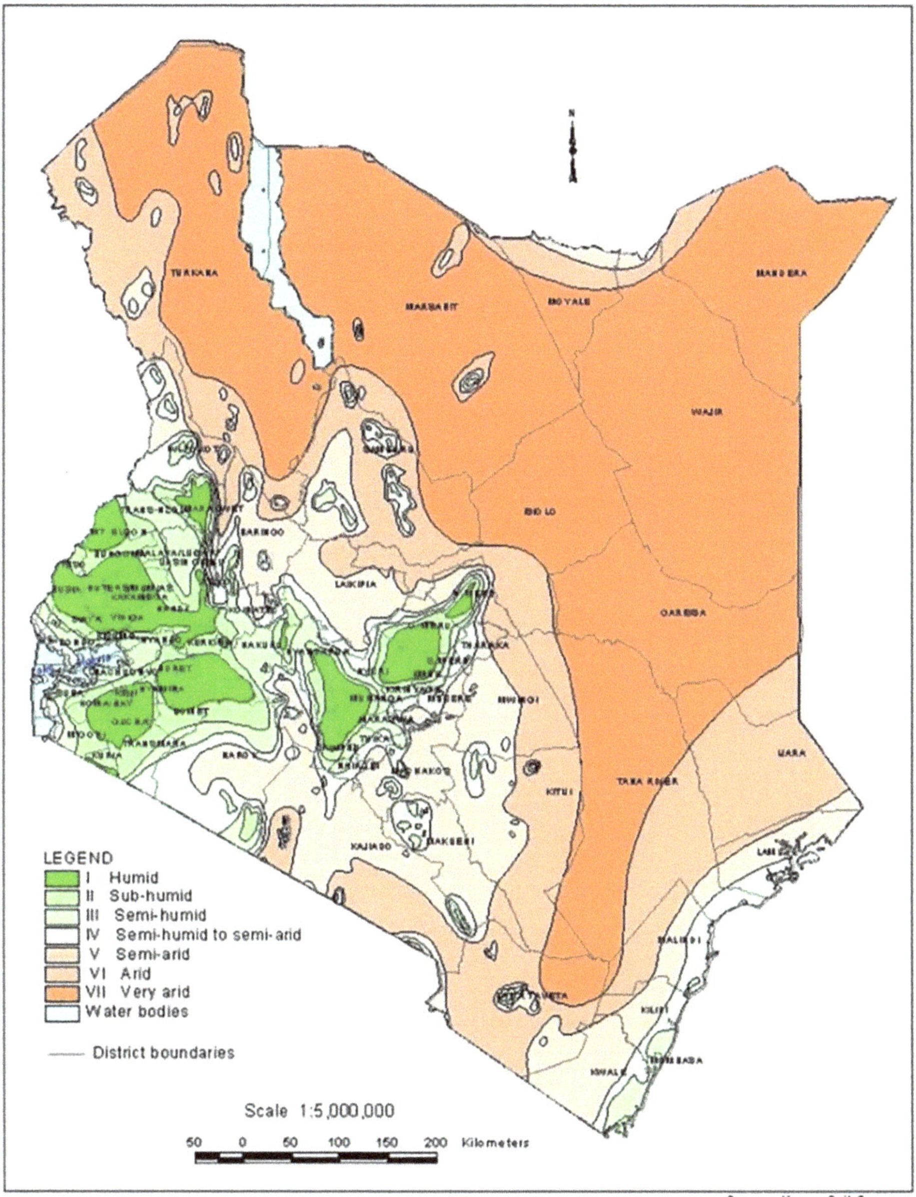
TURKANA
MARSABIT
MOYALE
MANDERA
WAJIR
ISIOLO
GARISSA
LAIKIPIA
KITUI
TANA RIVER
KAJIADO
LAMU
MALINDI
KILIFI
KWALE
LEGEND
I Humid
II Sub-humid
III Semi-humid
IV Semi-humid to semi-arid
V Semi-arid
VI Arid
VII Very arid
Water bodies
District boundaries
Scale 1:5,000,000
50 0 50 100 150 200 Kilometers

Source, Kenya Soil Survey

Figure 1: Agro-climatic map of Kenya by Kenya Soil Survey 1982.

Intensity of Agriculture

Agricultural Intensity is pronounced in the rainy highlands where temperatures are persistently cooler, evaporation rates lower and the soil has been blessed with rich volcanic nutrients. The highlands are divided by the Great Rift Valley which drops down to 1000 meters above sea level. Nairobi, the capital resides above 2000 meters in the Eastern Highlands which include the heights of Mt. Kenya at just fewer than 6000 meters. Commercial agriculture abounds with cultivation of tea, coffee and specialty export vegetables and flowers. Altitudinal zones are pronounced.

The Western Highlands occupy the high volcanic plateau (2500 meters) that slopes downward to Lake Victoria at about 1400 meters. Agriculture here serves more of the domestic market with the production of maize, wheat and livestock although export crops of tea and sugar are also prominent. Within the Rift Valley (Figure 2), the rainfall declines and irrigation frequently supports commercial agriculture of various types. Livestock herding is more pronounced in the Valley.

Figure 2: Looking into the Great Rift Valley. Potatoes in foreground over an intensively farmed bench terrace. Photo by Rick Bein 2011.

Traditional agriculture prevails throughout the country and is composed of crops that are grown mainly for home consumption. Many of these crops have long traditions of cultivation that carry much knowledge and lore about their uses. Generally these are high nutrition foods that are eaten at home or traded locally. Many of these crops along with numerous wild plants are also used medicinally. Some of staples of western Kenya include sorghum, finger millet , arrow root, beans, black night shade, sweet potatoes, kale, and cow peas (Figure 3). Maize cake (ugali) is still a major component of the diet. Many greens that grow semi-wild are consumed as vegetables. What might be considered a weed in other circumstances us frequently eaten as a salad.

Figure 3: Mixture of vegetables. Photo by Rick Bein 2011

The farmers of Kenya make many claims for the medicinal value of black night shade, products of the neim tree, sodom apples and the fats of sheep, goats, chickens, and pigs. One product called "black jack" is highly prized for its blood clotting capacity. Every house hold lists several unique medical plants. See figure 4.

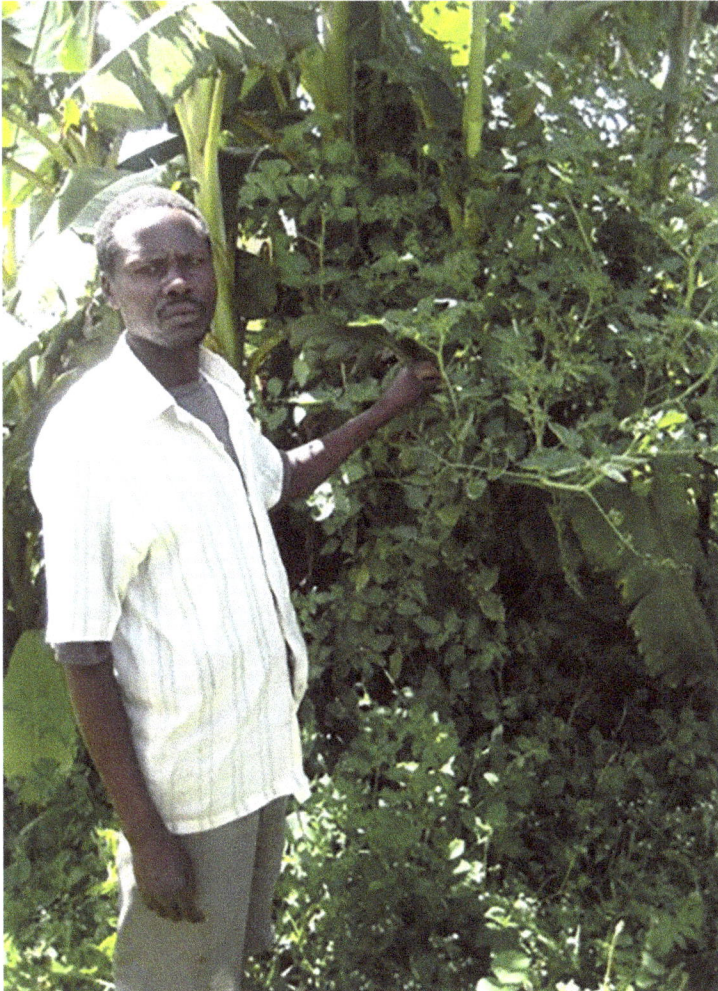

Figure 4: Medicinal plant. Photo by Rick Bein 2011.

Characteristics of Traditional Agriculture

One characteristic of traditional agriculture is the practice of growing the crops in a mix of crops and animals. This has been identified through out the tropics as polyculture, intercropping, agro-forestry and multi-story agriculture. See figure 5. The system provides food security as plants are maturing through out the year; the different crops have symbiotic relationships with each other providing pest control, trellises, shade and fertilizer.

Figure 5: Mix of grasses, pumpkins and pigeon peas. Photo by Gilbert Nduru 2011.

In recent years there has been increase in demand for different crops in Kenya. This has been a result of frequent crop failures and declining yields of the main staple crops of maize, beans and potatoes. More farmers are turning to growing traditional food crops such as millet, sorghum (figure 6), bananas, yams, greens and sweet potatoes. These are now becoming cash crops and are appearing more often in local markets and in restaurants.

Figure 6: Sorghum is seen as a nutritious, drought resistant food staple. Photo by Gilbert Nduru 2011.

Seasonal Variations

The landscape can appear dramatically different depending on the season (Figure 7). Located on the equator there is no winter and seasonality is defined between "wet season" and "dry season". Although these seasons are expected to occur at certain times, that frequently varies and it is most noted when the rainy season arrives late or is prolonged.

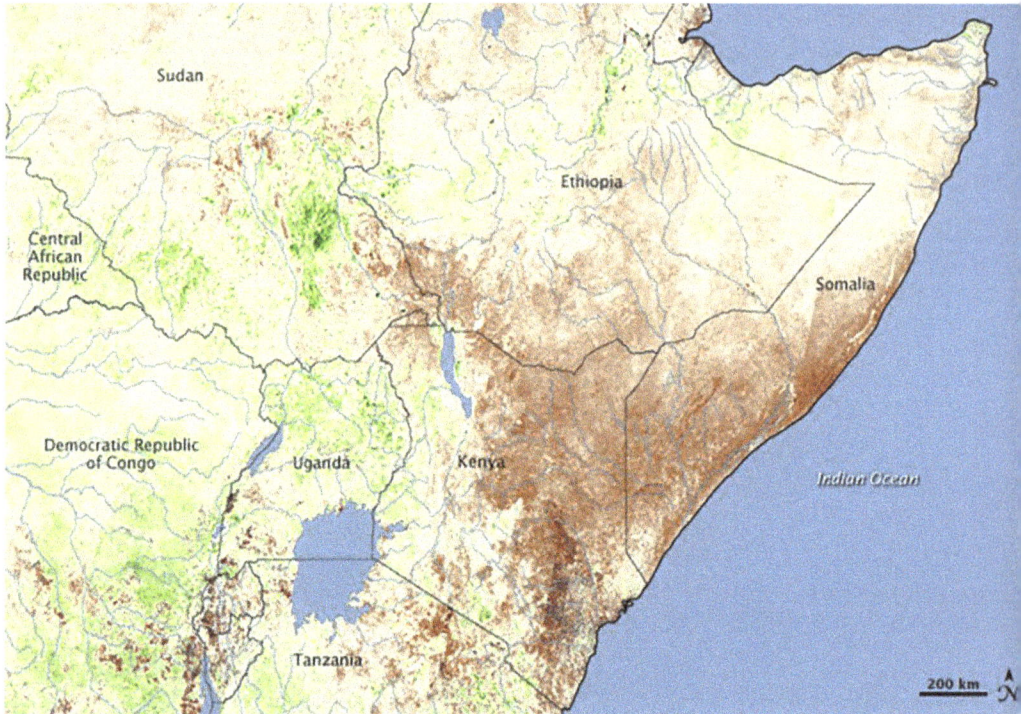

Figure 7: French Spot image shows the intensity of the dry season January 2011 The brown color (dry lands) dominates much of Eastern Kenya.

Impacts of Climate Change

Climatic variability is normally associated with shifts in rainfall patterns. This involves reduced amount of rainfall at a time when farming folks expect rains. Droughts produce crop failure, food shortage and lack of pasture land. People will be forced to leave their land in search of water and food. Excessive rainfall when its least expected results in Flooding and Destruction of property.

Changes in rainfall and temperature can have an effect on the Kenyan agricultural base. Most serious would be increasing temperatures which will impact dramatically the small elevation zones and force the adaptive agriculture up slope. Declining precipitation would impact the conventional food staple producing areas (maize and beans). The impacts would vary from zone to zone.

Conclusion

The advantages of traditional agriculture are multiple. Different crops mature at different times, ensuring food security during the year. Intercropping/polyculture farming practice reduces labor costs with regard to weeding and pest control while the different crops support each other with fertilization, shade and pest repellents. Wild products are available at different times of the year in the form of plants, roots, legumes and tubers. Some crop products and plants are medicinal. Most of the traditional crops are well suited to harsh climatic conditions currently prevalent over most parts of Kenya.

More study is required on the role of traditional food crops in ensuring sustainable food production and supply in Kenya and other parts of Africa. More research is necessary on the role traditional crops as sources of income. There is need for more research on the nutritional value of respective crops. There is a need for a framework for the promotion and protection of traditional farming that should involve farmers in policy development and marketing.

20. ASPECTS OF SUDAN IN MID 1970's

Aspects of Khartoum in the mid 1970's

The Sudanese people are some the most genteel people on earth. As a white foreigner, I was given great respect and kindness. (Being white, I was given undeserved preferential treatment). I was welcomed into their homes and given tours of their towns and farms. I felt that I was given kindness that exceeded what they gave their own people. Being in Sudan was one of the joys of my life. I arrived with my family in late in the summer of 1974 and enjoyed three years teaching and researching at the University of Khartoum Geography Department.

Most of the Sudanese are practicing Muslims and I became quite comfortable being around their devout practices. Six times per day, we would stop for a few minutes so the devout could pray toward Mecca. The Sudanese Muslims are very loving people and do not judge others who are not Islamic. They did condemn those who used Islam as a cover to justify violence. There was a small Coptic community in Khartoum, and a majority of Christian sects in the far south as well as tribal religions in remote areas of the country.

The lives of the Sudanese were not easy as they endured many hardships: floods, droughts, dust storms, food shortages and periodic military coups. Prior to my arrival the most recent coup was led by Jaafar Muhammad al-Nimeiry who took over the government in 1969 to be dictator for 16 years.

Map of the Nile valley by CIA.

In 1955 the British colonial masters left the Sudanese with a democratic self-rule system that lasted for three years, enough time for the people to become appreciative of the democratic process and were distressed when a military coup occurred 1958. This autocratic rule lasted 6 years with much resistance until another attempt at democracy began in 1965. Four years later Numeri took over and was in place when I arrived.

During my three years stay in Sudan (1974-77), I served as a lecturer at the University of Khartoum Geography Department where I heard a lot of discussion about their hopes of reinstating democracy. This expression was allowed on the campus where the free-thinking mission of the University was permitted. It was considered somewhat safe expression if it stayed on campus. Although dissident Sudanese still sent their sons and daughters to the University of Khartoum with the idea, they might instigate a revolution.

https://www.stepmap.com/map/khartoum-YM7I9Z043R

Several military coup attempts occurred while I was there, but none of them were instigated at the University. Even then, when the attempts happened elsewhere in the country the University was barricaded by the army because of its pro-democracy sentiment.

Keeping up with these coup attempts was no easy task. Once while I was driving to the campus with my wife and three year old twins, we were surprised by a military tank heading straight for us. Immediately I told the twins to get on the floor between the seats (Amazingly they obeyed instantly with no whining) and did a quick "U" turn and headed back to cross over the Blue Nile Bridge. I was fearfully grimacing, thinking that at any moment an exploding shell would do us in. Lucky for us nothing happened. But the tank operator probably did not see us a threat or else did not have enough ammunition to waste on a family vehicle. Whatever the case, the coup failed.

On the next coup attempt, we were driving to campus again and as we neared our destination the twins suddenly started crying because their eyes hurt. Then we noticed it too. "Tear Gas!" the military was trying to quell any action from the students. We wiped our eyes and just headed back home again to learn from the neighbors that there was a coup attempt.

That was a good lesson. "Don't go anywhere without checking with the neighbors". They always seemed to know what was going on. One major clue about a pending military coup was that the radio and telephones would go dead. The first thing the rebels would do was to cut off any communication to keep the government from organizing against them. Although my Arabic was limited, with no sound coming out of the radio it did not take much to figure out when something was amiss.

Another coup attempt occurred when Muammar Gaddafi of Libya sent a military contingent to Khartoum to overtake Numeri. We had taken students on a field trip down to the 6th cataract and were returning when our bus driver noticed all the traffic was going the other way. Something was wrong in Khartoum! Our driver heard about if from drivers going the other way, but we proceeded cautiously on to Khartoum North where the under equipped Sudanese army had cordoned off access to the Nile Bridges. We could stay in Khartoum North where it was safe. We dropped the students off at the agriculture campus and the faculty stayed at my place. (Fortunately, my family was on vacation leave back in the States).

Since food was forbidden to cross the river into Khartoum proper, we had plenty to eat in Khartoum North. After a week Gaddafi's starving solders were easily overcome. The people living in the main city had sequestered themselves and refused to share their food stashes with the invaders. Gaddafi's solders expected the Sudanese locals would welcome them with open arms. That did not happen as people said, "We may have a dictator, but at least he is ours".

Eight years after leaving Khartoum, I had a chance to return for a short visit in 1985. I went to visit the Geography Department at the University and arranged for a tour to the Gezira agriculture scheme located between the Blue and White Niles. Upon return to the city, I went directly to the airport to catch my flight out. As I arrived at the gate there was news of a military coup taking place and they were planning to close the airport. Fortunately, my plane was the last one out.

This military coup was successful and over-threw Numeri's 16-year reign. After a series of coups, the new dictator Omar Bashir emerged and reigned until 2019. As a very harsh ruler, a massive persistent movement of civilians forced him to step down. Now there is a transition military-civilian coalition slowing developing a democratic system.

As I arrived to Sudan in 1974 the **Geography Department** welcomed me warmly and assigned me to teach Environmental Conservation and the Geography of Agriculture. I had just received my PhD at the University of Florida and I was up for the adventure. The University of Khartoum had kindly provided transportation from the States and arranged our living arrangements.

The house where we lived was in Khartoum North across the Blue Nile. It was in a neighborhood called Khobar, a developing area where new houses were being built. An eight-foot brick wall with an iron gate surrounded the house that contained spacious ten-foot-high ceilings and a stairway going to a flat roof. The design was arranged to deal with the extremely hot climate, 100 degrees Fahrenheit plus in summer. The high ceilings trapped air so that the warmer air would rise leaving slightly cooler air on floor. The roof was

Dust storm approaching Khartoum. Photo by Rick Bein 1976.

There was an interesting environmental problem that I had not encountered before. Most of the soil outside of the Nile flood plain is a black-cracking clay that shrinks when it is dry leaving big cracks and expands when wet, closing the cracks. This means that any structures built on it would be impacted by this shifting soil. One way that my landlord dealt with the problem was to keep the soil dry as much as possible and in this arid region rain was not a serious issue. One time when we thought we could grow some vegetables in the compound next to the house, he immediately pulled up all of our plants because the act of watering them would cause the ground to swell.

Black clay swells when wet and shrinks when dry. Photo by Rick Bein 1976.

The university students were very bright as they had been selected at several stages of advancement from grammar school, middle school and high school to be admitted to the only University in the country. They also had to pass an English competency exam since all university subjects were taught in English. That changed some years later to Arabic.

It seems that I broke one of the rules that old British academics had left behind and that was that faculty do not fraternize with the students. Students were held at a distance and were treated rather rudely, it was like a rite of passage, where those who could endure, the humiliation were allowed to graduate.

I worked closely with the students, took them on fields trips, and visited their homes. They explained a lot of their traditions and they tried to teach me Arabic. Fortunately, I was not ostracized for this, as I felt that this was a great way to learn the culture. By chance one of my students later earned his PhD and came to teach Geography at Hunter University in New York City. I ran in to Mohamed Ibrahim at one of the Association of American Geography conferences. He shared with me that my teaching inspired him to become a teacher because I alone at the Khartoum Geography Department, "treated students like human beings". We still get together at the AAG meetings every year.

My research interest was in traditional agriculture, and it was waiting for me in Sudan. Agriculture was the mainstay of the economy and there was plenty to study. I was intrigued with the land use patterns and the people's relationship with nature. There was farming on the banks of the Blue Nile that passed right by campus. This farming, called "Gerif" is probably the oldest form of agriculture along the Nile and extends all the way through Egypt. Gerif cultivates the silts that are exposed when the annual Nile flood waters recede. Crops are planted one row at a time following the descent of the water.

The gerif is cultivated every year as the Nile flood waters recede. Vegetables on the left are ready to harvest while plants on the right are being planted. Photo by Rick Bein 1976.

An additional type of farming developed in cultivating the entire floodplain that was productive enough to support the ancient Egyptian civilizations beginning four thousand years ago. It is the high flood that happens about every six years or so, when extra heavy rains in Ethiopia send enough water to immerse the Nile floodplain depositing its precious silts.

The Nile floodplain is covered with silt (in fore ground) when the high flood occurs. The Nile channel is below the picture. The village of Al Uqsur, Egypt is in the desert out of the flood plain. This image precedes the construction of the Aswan Dam holding back the high flood and its silt deposition. Therefor no more silt can be deposited to restore fertility. Photographer unknown

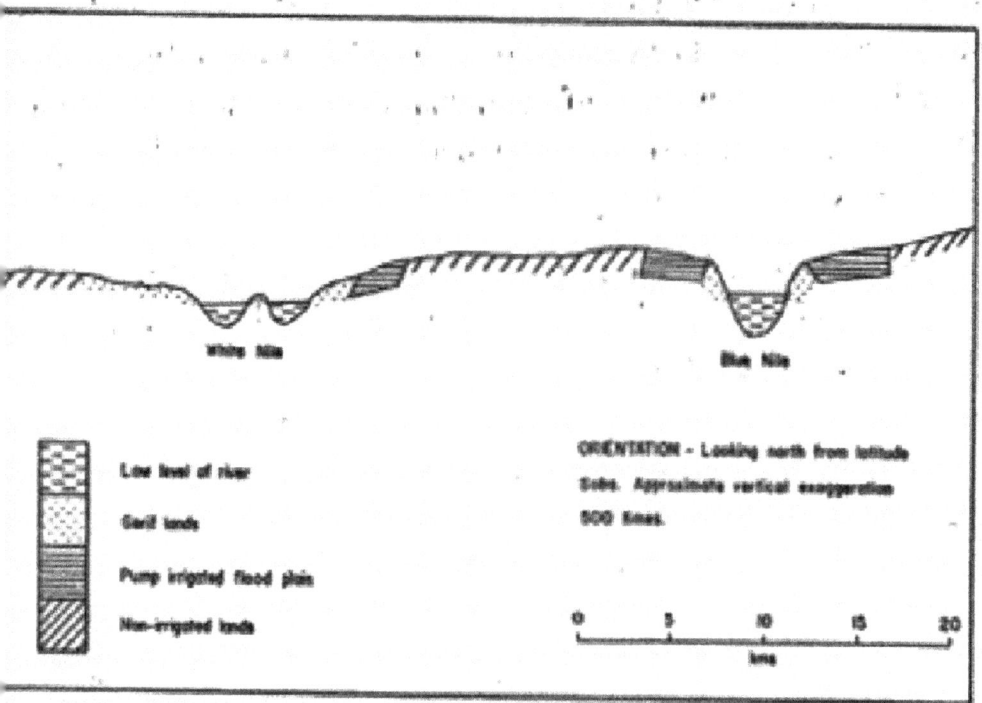

Three types of agriculture plots are shown with the Gerif farming along the sloping banks of the river, shadowed by the irrigated floodplain, and the expanded farming beyond the floodplain. Image from Bein, Frederick L. (1978) "Land Use Patterns along the Nile." Sudan Notes and Records. Vol. 58, pp. 180-189.

The construction of the Aswan Dam drastically changed this flood plain agriculture in Egypt since the rejuvenating silts no longer enrich the flood plain in Egypt. Diesel fuel pumps now irrigate the flood plain as well as larger farms in the desert. Chemicals are now applied to fertilize the land. However in Sudan, upstream from the dam, the high flood still occurs, depositing its silts on the flood plain.

Water is pumped up onto the flood plain to irrigate the crops. Ancient water lifting devices, the shaduf (water wheel) and the saquia (lever) were still in operation in Sudan in the 1970's where diesel fuel pumps could not be afforded.

A Tuti Island farmer harvests kale to send across the river to the Khartoum market. Photo by Rick Bein 1974.

Specific crops tend to change with distance from the Khartoum market. Perishable vegetables are cultivated within a short distance from Khartoum. More durable fruits and vegetables are grown further away. See land use map below).

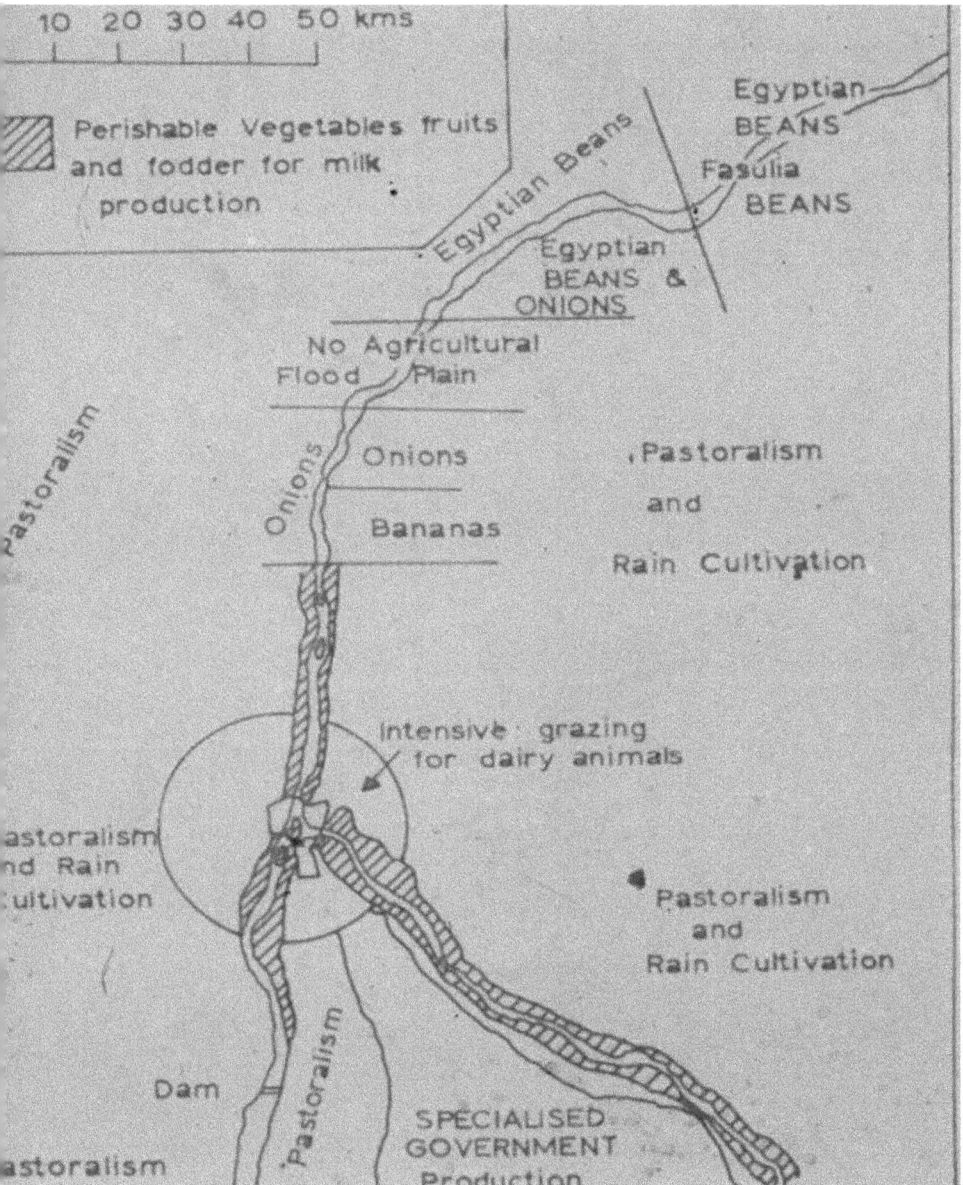

The Land-use map on the right depicts different crops growing along the river. Image by Bein, Frederick L., (1978) "Land Use Patterns Along the Nile." Sudan Notes and Records. Vol. 58, pp. 180-189. Photo by Rick Bein 1976

Wad el Basal, beyond the edge of the date trees that line the flood plain, a Faqui (holy man) pumps water into the desert where he runs a Koranic school where boys between 3 and 8 years learn how to raise onions. At 8 years of age most of the boys leave and start primary school. Two of my university students are pictured on the left.

The ancient arrangement provides each property with frontage on the Nile, where farmers cultivate the gerif and the irrigation floodplain. Ownership is inherited and divided among the heirs and the narrow piece of land is sliced even smaller. Title to the land can be held by hundreds of heirs, but one relative does the farming by agreement. Today, urbanization had offered much more opportunities where all the other heirs can occupy themselves.

Mechanized agriculture Flood plain Gerif

Three types of traditional long-lot farms line the Nile are defined by the notations under where they occur. Image photographed from commercial airline by Rick Bein 1976.

A diesel fuel pump takes water up beyond the floodplain to irrigated crops in the desert. Photo by Rick Bein 1976.

The Gezira scheme was started by the British early in 1925 to grow cotton for the English textile mills. A million acres were brought under irrigation from the Senar Dam on the Blue Nile. Water from the reservoir was distributed through a 2,700-mile network of canals to the farms between the Blue and White Niles. Following independence in 1952, farmers now grow many other crops and send some cotton to Sudanese textile mills. Other crops include oil-seeds, peanuts, wheat, sesame, sorghum, millet, and vegetables.

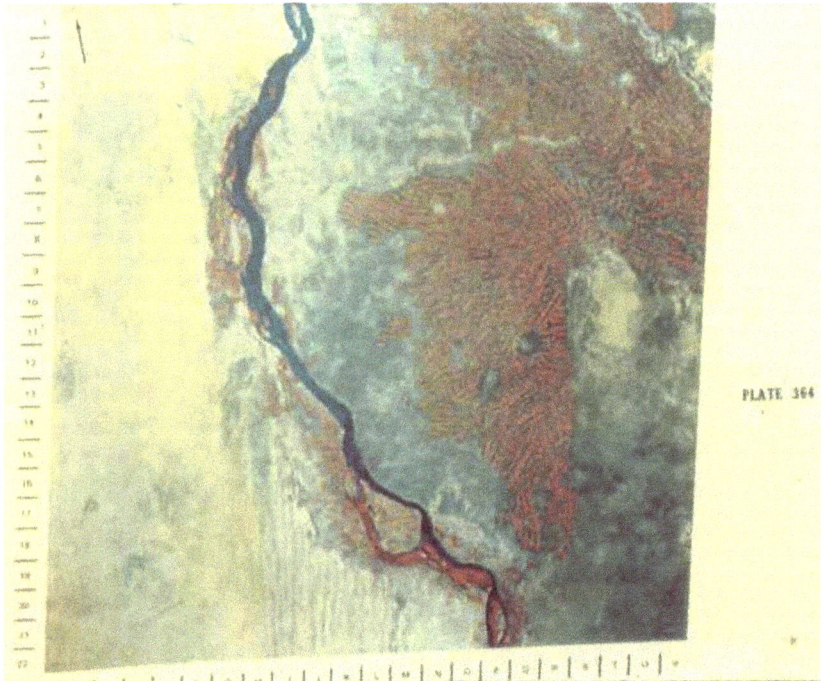

PLATE 364

This Infrared image of part of the Gezira scheme shows the White Nile on the left and the Blue on the top right. The farmland is shown as red. Image from Landsat.

Sorghum is the crop most cultivated in Sudan and is dependent on rainfall. In the semi-arid Sahel away from the Nile successful crop years are not that frequent. Maybe two good years in a five year period provide enough grain to eat. When there is good rainfall year, more than adequate grain is produced. The extra grain is stored in pits (matmuro) on the periphery of villages where it will remain hidden until needed in a few years. About a quarter ton of grain is stored in each matmuro and covered with a layer of chaff from the previous harvest and a layer of clay soil is added on top. This grain remains edible for up to seven years. Each farmer knows exactly where his grain is hidden in this seemingly endless clay plain. This unique practice supports the village in good and bad years, remaining undisturbed by any marauding army or other tribes. If the villagers need to flee they know when they come back in a few years there will be sorghum waiting for them.

Matmuro grain storage preserves the grain for several years. The black clay swells when wet and seals the grain from any rainfall moisture. The chaff keeps the crumbs of the soil from mixing with the grain and can later be easily winnowed out. Photo by Rick Bein 1975.

This rain fed agriculture extends four hundred miles to the south where the Nuba Mountain people also maintain some unique farming traditions of the Sahel. On the flat land sorghum is grown in much the same manner. The only addition is the women have their agriculture apart from the men. The women farm the steep hillsides by using terraces (Jebraca) to reduce erosion.

The women farm the steep hillsides by using terraces (Jebraca) to create more level places to plant sorghum. Photo by Rick Bein 1977.

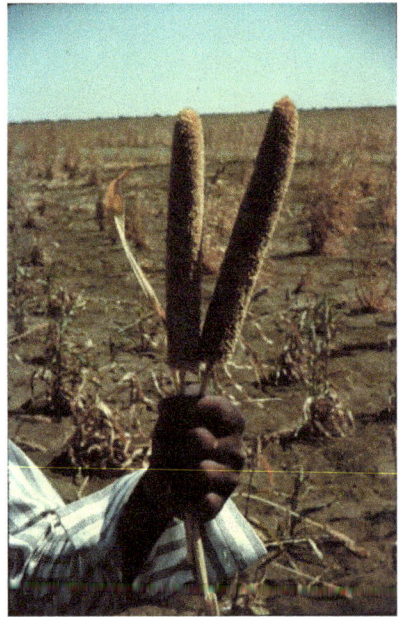

Nuba grain storage structure for sorghum. Sorghum seed for the following year, hangs on outside of the structure. Millet heads, Images by Rick Bein 1977

Traveling within Sudan provided a more extensive perspective of life throughout the country. My first visit was to Dongola, north of Khartoum along the Nile. An American colleague, Chris Winters, went with me. This trip took two phases. First, we took a convoy truck (lorries) at night across the desert to Merowe. Traveling at night was advantageous as the lorries spread out about a quarter of a mile apart to avoid each other's dust. The lead lorry navigated with the stars, while the others followed the taillights of the lorry ahead. We arrived in Merowe at dawn and spent the day visiting ancient ruins and on the following day boarded a steamboat that took us down the river to Dongola.

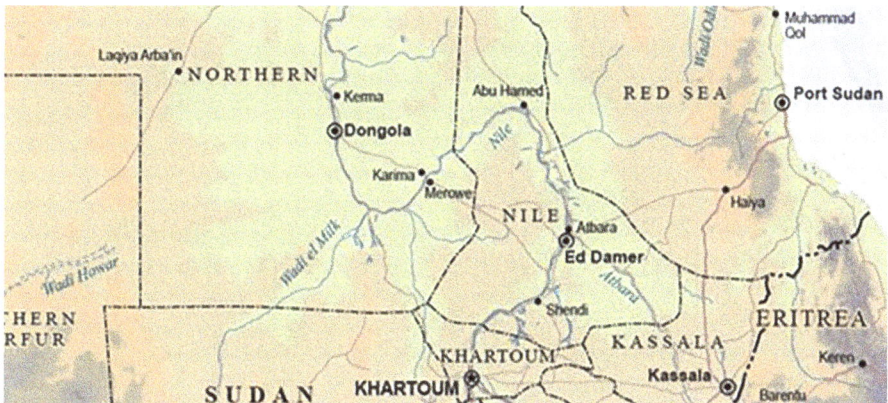

https://joshuaproject.net/people_groups/11607

Passenger lorry by Rick Bein 1975. River boat by Creative Commons CC0 (CC0 1.0 Universal (CC0 1.0) Public Domain.

One embarrassing memory near the beginning of this journey occurred shortly after departing Khartoum. The Lorry left the city at four PM and drove until sundown where we stopped, and everyone descended. A blanket was spread on the ground where food donations were laid. This was the Ramadan breakfast which occurs when the sun sets after a long day of fasting. Muslims consume no food during the daylight hours for the month of Ramadan.

A wealth of food is brought out to share. Only the right hand is used when eating the meals. Picture by Rick Bein 1975.

Although we had no food to share, they invited us to share in their feast. The food was delicious, and everyone picked food up with their right hand. At one point I negligently reached for some food with my left hand, something that is never done in Islamic society. The people began leaving the circle, disgusted at what I had done, particularly at the sacred Ramadan breakfast. I was mortified. Later they forgave me for my mistake.

You may have noticed above that there are no women in the circle of desert diners. The women get to eat what the men picked through.

Another eating event occurred after we arrived in Dongola. At another evening breakfast in a small neighborhood a hors d'oeuvre was brought out while the main meal was being prepared. Marara, eaten traditionally in many parts of Sudan, consists of a plate full of raw sheep entrails. As the guest I was asked to serve myself first. I was aghast at this presentation, let alone eating any of it. I looked to ask my colleague, Chris, what to do in this situation, but he had suddenly disappeared. Being so new to the culture I was unsure what to do and did not want to offend anyone again. I decided that I had to be a good guest and eat. So, I chose a small piece and stuck in my mouth and began to chew. Even though being small, it was difficult to manage because it was like chewing on a small piece of rubber hose. Mixed with my saliva it began to swell and when I bit down hard, some liquid squirted out into the back of my mouth! To avoid retching I just gulped and swallowed it down. Having seen my accomplishment, they were delighted to offer me another piece! I declined since I had already done my duty.

Later, back in Khartoum, I related this story to some of my Sudanese friends who all gagged and said we never eat that stuff! No one would have been offended if you had refused.

Professor Mustafa Khogali was studying water resources along the Atbara River in the Butana Desert northeast of Khartoum and invited me to go with him. The Atbara River is a little-known tributary to the Nile that contributes about 10% of the water during the rainy season. During the dry season, the time of our visit, the river dries up leaving stagnant pools inside oxbows.

Water remains pooled in an oxbow of Atbara River during the dry season. This water is polluted with animal and human waste, and with shistosomes. Nomads are camped nearby while their children play in the polluted water. A shallow hole is dug about ten inches from the edge of the standing water allowing water to seep through, filtering out some of the contaminants to produce a more potable water. As the water evaporates another shallow hole is dug closer to the retreating water. Photos by Rick Bein 1974.

We continued our journey further out to where no ground water was available. We came upon a herd of camels harvesting the succulent berries from thorn bushes. Camels can get their water from these tiny fruits. No other grazing animals can survive here. When we stopped to take pictures, a nomad came running out of the bushes and asking if we had any water.

"Sure" we said and pulled out a jug of water and poured him a glassful. He drank that up in a hurry and after two more glassfuls he had his fill and thanked us. When I asked him

"When was your last drink of water?"

"Six weeks ago,"

I asked, "How do you survive so long without water?"

"Camel's milk, he replied."

He showed us a little leather bag where kept some camel's milk. I smelled it and it had a rancid odor but had to taste it. Its taste exceeded the odor, and I did not swallow it. Everything was fine for a while after that. But the next morning I was miserable and for two more days I still had to keep an eye open for restrooms or bushes to hide behind.

Bedouin and his camels. Notice that the grass has not been grazed, meaning no cattle, sheep or goats can survive here for lack of water. The agile lips of the camels can reach between the thorns and pluck the tiny fruits that sate the camels' thirst. The camels are being fattened to be sold as meat in Egypt and Port Sudan. The nomad stands beside his herd camel that he rides to drive the others around. She also shares her milk with him. Photos by Rick Bein 1974.

After a year in Sudan, my Arabic was adequate to travel by myself. One trip I took was to Darfur in western Sudan. I took one of the transport lorries to the town of El Obeid, Kordofan province and transferred to another lorry to Genena close to the Chad border. We stopped to let people off at villages and I descended to take some photos.

Thorn bushes are collected and used as fences to keep out domestic animals and pests. Photo by Rick Bein

Children guard harvested sorghum from birds. Photos by Rick Bein 1975,

When we left, the village dogs came barking after us for several miles. Every village had a pack of dogs that would emerge and chase any vehicle that passed. I was impressed with the speed and endurance of these dogs. I believe they could compete well with racing gray hounds. It seems that the dogs do not belong to any one villager, but everyone gives them food scraps. Children play with them, but their main function is to contain the village rats.

Darfur (home of the Fur) was an independent sultanate for several hundred years and was incorporated into Sudan by Anglo-Egyptian forces in 1916. Being 600 miles from Khartoum and the culture of the Nile, the Fur have very little affinity for the rest of Sudan. Not being Arabs they have operated somewhat independently but in recent years when they tried to take their independence, Omer Bahir's military forces squashed the attempt. A semi-vigilante Arab group called the Janjaweed, sponsored by Khartoum government pillaged and destroyed villages and disrupted their water supplies. Because of this war between Sudanese government forces and the indigenous population, Darfur has been in a state of humanitarian emergency since 2003.

Polygamous family building a house in Genena, Darfur. The Koran allows a man to have four wives, an arrangement which brings security for the women and wealth for the man.

Men ride on top of the train for free while they buy tickets for their wives and children to ride inside. Photos by Rick Bein 1975.

My visit there in 1976 was met by peace and tranquilly and I spent some days meeting people taking little tours to nearby villages. As I was walking around Genena, I saw a Fur family building a house and from a shop across the street I was able to photograph their activity. What I saw was the man and his son up on some homemade scaffolding building a wall with bricks. He had four wives helping down below. One was mixing hod while another carried it to a third wife who relayed it up a ladder to the brick layers above. Meanwhile the fourth wife showed with a large tray of food for the midday meal.

Another occurrence was meeting Assam a young man who had earned his bachelor,s degree from the Geography Department at the University of Khartoum. He had been working in a government office for a few years. As we talked, he expressed his interest in returning to the University to pursue a master's degree. I encouraged him to do so and I carried copies of some of his documents with me back to Khartoum. A month later he showed up on campus and within a year he had finished his master's degree and headed out of the country to complete a PhD. The last I heard he was a professor of Geography at King Saud University in Saudi Arabia.

While still in Genena, Assam told me of a more comfortable alternative way to return to Khartoum. There was a train that ran from a neighboring Nyala, a town about 60 miles away, on to Khartoum. I decided that would be another adventure and when I left Genena, I took a lorry to Nyala where I bought a ticket on the train. That was an interesting journey. I noticed that there were only women and children in my train car. Once we pulled out of the station, a crowd of men (probably the husbands) came running up to the train and climbed the ladders to the roof. Fortunately, the train could only travel at 15 miles per hour because of the poor condition of the rails. It was not too dangerous for a free ride! At the consternation of the porter in my car I decided climb up there also, but just long enough to take this picture.

I had heard about the Nuba Mountain People, a group of tribes unique to themselves living in the province of Southern Kordofan about 400 miles southwest of Khartoum. I decided to go visit the Nuba in 1975 and took a passenger lorry to Kadugli the provincial capital. In Kadugli I began looking for a vehicle going to the Nuba Mountains. I ran across a young couple who had visited the village of Buhram a few months prior. They had driven their camper to the small community and were impressed with their visit. They suggested that I go there since I had no other plan. I began asking around the market for a lorry going there. I immediately found one leaving first thing in the morning.

- The mountains cover an area roughly 64 km wide by 145 km long (40 by 90 miles), and are 450 to 900 meters (1,500 to 3,000 feet) higher in elevation than the surrounding plain.

- The mountains stretch for some 48,000 square kilometers (19,000 square miles).

- The climate is semi-arid with under 800 mm of rain per year on average, but lush and green compared with most nearby areas.

- There are almost no roads in the Nuba Mountains; most villages there are connected by ancient paths that cannot be reached by motor vehicles.

- The rainy season extends from mid-May to mid-October, and annual rainfall ranges from 400 to 800 millimeters (16.4 to 32.8 in), allowing grazing and seasonal rain-fed agriculture.

- Population is approximately 1.7 million in 2019

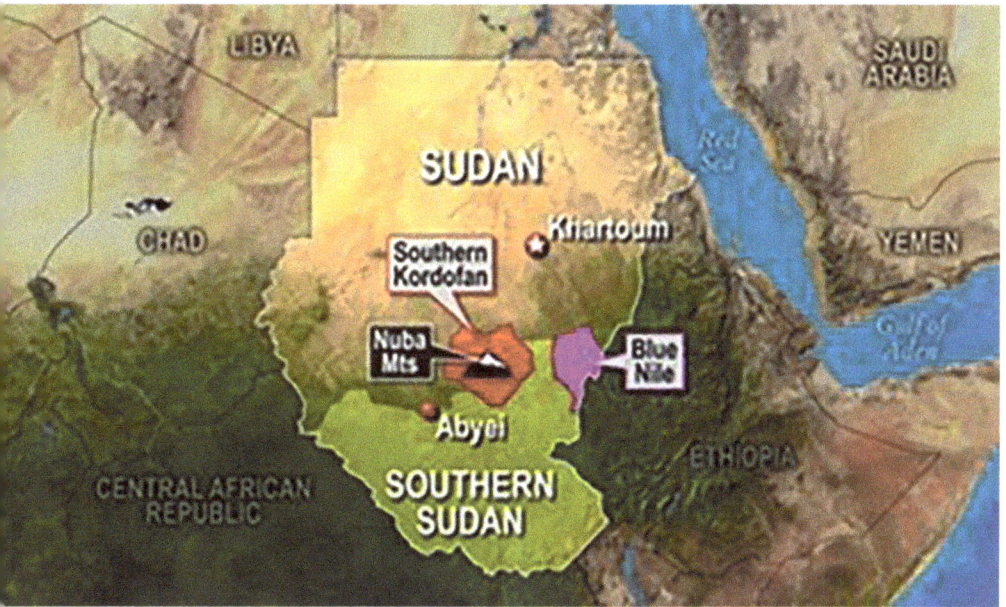

Statistical summary of Nuba Mountain

The lorry was carrying household goods and six passengers including me. As we approached a craggy outcrop emerged on the horizon and we proceeded directly toward it. At its base was an Arab settlement of about 4 small shops the homes for about 10 families who greeted me warmly and wanted to know why I was there. I told them that I came to see the Nuba. They told me to go around the base of hill until you see them. The Arabs told me that when I was finished, I should return to them because they had a place for me to sleep and I would join in a feast that they were preparing. Not knowing what was ahead, I accepted their hospitality.

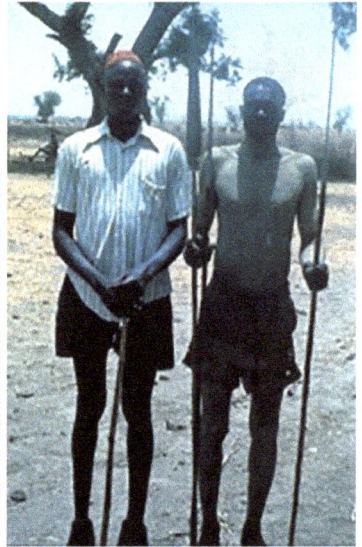

Approaching Buhram community showing local Arab brick works, and cattle nomads. The lead Nuba farmer is wearing my extra shirt, skull cap boots and shorts. Also included is the local metal worker showing his spears made from scraps left from Rommel's WW2 military hardware. Rick Bein 1977.

I grabbed my backpack and after 300 yards I came upon a few Nuba people we were happily curious about me. A few of them spoke a few words in Arabic and I told them I wanted to see their agriculture. They took me over to sit with them around a tree. The dressed only in shorts, men and women alike. Everyone was barefoot and wore nothing on their heads. They were all large people, no one under 5'10". Several of the Nuba came to look at this strange white man; they seemed as curious about me as I was about them. Some women bought some sorghum mash to eat and some water. I really was hungry and thirsty. I knew that the water might not settle well with me, but I wasn't going to run back to the Arab settlement just for a drink.

Meanwhile, they brought someone they all recognized as being one of their better farmers. They wanted to address my mission to visit their farming areas. He decided to match his dress as closely as could to mine. He had a skull cap and wore probably the only pair of boots in the community. He seemed somewhat uncomfortable in the boots. We started out, neither of us able to communicate verbally. He pointed out their main crops, sorghum, and millet which I was familiar with. Then we came to a field of 3-foot stalks standing about 2 feet apart. What are these? He said "gooton" which was the

Arabic for cotton. The stalks caught me off guard since in the States cotton farmers get rid of the stalks since they might carry diseases that would affect next year's crop.

Cotton Stalks left behind after harvest. Nuba woman passing up drinking water colored by Kaolin. Photos by Rick Bein 1977.

This was a puzzle to me. They did not have any capability to process cotton. "Why cotton?" "Groosh" was his reply. Arabic for money. What do you do with Groosh? He showed me a plug of chewing tobacco from his pocket.

Then the relationship between the Arabs and the Nuba became clear. The Arabs played to the Nuba addiction to chewing tobacco which they supported by trading their cotton. When we got back to the tree, I took notice that the whole community was hooked on chewing tobacco. These Arabs being Muslim also had a mission to recruit the Nuba to Islam. A few Nuba had converted.

Instead of going back to the Arab settlement, I became fascinated by the sorghum grain threshing that was happening.

Nuba Women going out to the underground granaries to bring sorghum heads back to the village to be threshed. On the right, men and women cover their bodies with sacred flour and beat the grain heads with threshing boards. In the bottom right corner sorghum heads are ready to be threshed.

Sorghum heads waiting to be threshed in the for-ground, and immediately above the already threshed grain that will be stored in a matmuro. Woman winnows some grain before being ground into flour. Photos by Rick Bein 1977.

While I was taking pictures of all the happenings, the village architect came and said "come with me to see the houses". He was very proud of his artistic ability and wanted to show off the homesteads that he had completed. Each family lives in a small compound made of a circle of five round structures. Small mud walls connect each of the small buildings to create a closed circle. Each small building has a different purpose. The most important is the granary, holding the food supply. It is decorated with elaborate designs to keep away evil spirits. The keyhole shaped door provides the entry way which is also the master bedroom. There is a separate room for cooking and food preparation. The other two rooms are for the children. This housing arrangement still continues today and can be seen from expanded "Google Earth" views.

144

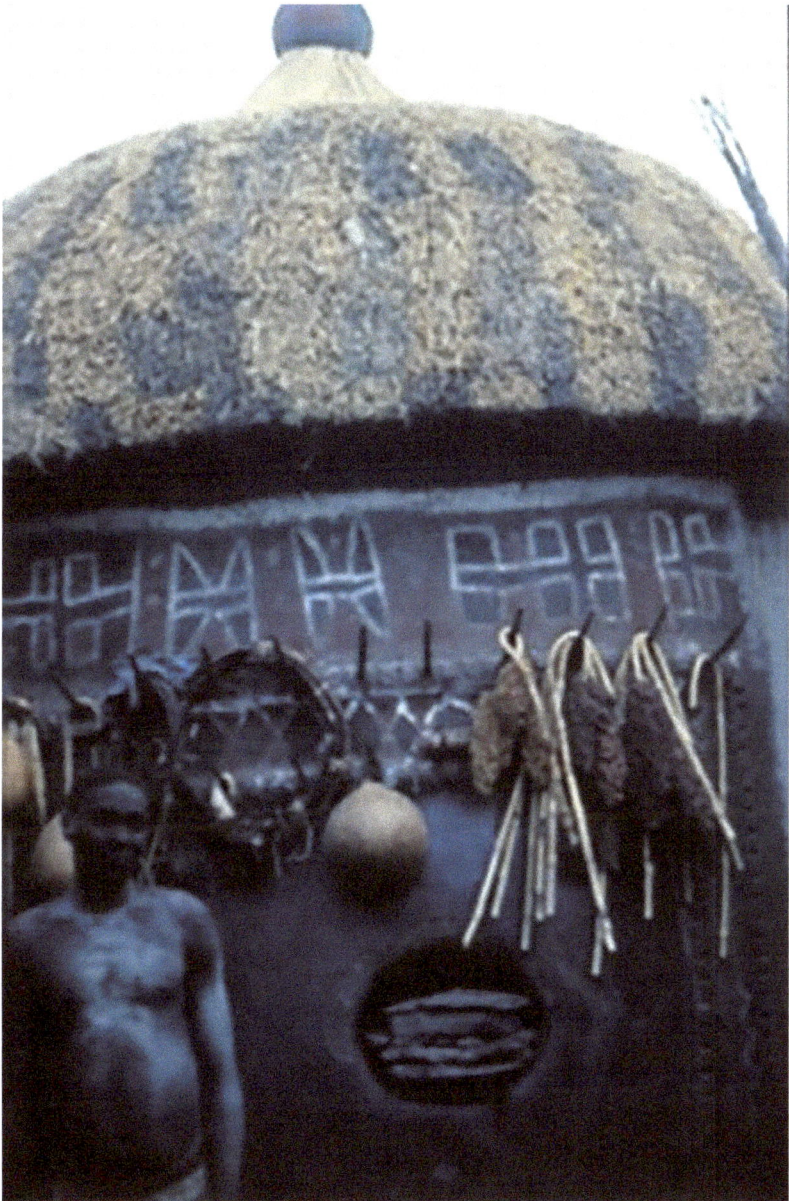

The master Architect shows off the household granary. His home sits on the hill to provide a lookout for enemies and to see arriving supply trucks. Photos by Rick Bein 1977.

The master Architect shows off the household granary. His home sits on the hill to provide a lookout for enemies and to see arriving supply trucks.

Photos by Rick Bein 1977.

Among the many other cultural attributes, I choose to focus on two items of interest. The Nuba are known for their Martial Art of wrestling. Children learn to wrestle in their play time. This continues and becomes a rite of passage when reaching adulthood. Women as well as men are excellent wrestlers. On certain evenings just before sundown teenagers compete in wrestling matches. Adults maintain their skills should they need to protect themselves from invaders. The wrestling matches consist of the combatants posturing (more pronounced for the girls) trying to intimidate their opponent, but the actual contact is extremely quick, usually thirty seconds. The winner merely must throw the other on the ground.

I spent a bit of time with the village champion who half in jest challenged me to a bout. Although we were the same size, I decided I would not stand a chance and declined. Anyway, we both had a good laugh. In a prior time, the Nuba and their martial arts were recruited to serve in the Sudanese army. Ironically with the reign of Omar Bashir (1985-2019, the Nuba suffered series Genocidal attacks and were forced to live hidden in caves.

The baobab tree bark serves as a source of cloth. It is beaten until soft, and the sheets are fashioned into shorts. This tradition had diminished greatly when I was there because western style clothes came free from the developed countries. Unfortunately, Goodwill does not provide feminine products and the women used the baobab fiber.

The entire community, Arabs and nomads come to watch the excitement. The baobab tree bark serves as a source of cloth. It is beaten until soft, and the sheets can be fashioned into clothing. Photos by Rick Bein 1977.

The Nuba community seemed enamored by my willingness to participate in whatever they were doing, that they forgot that I came to see their agriculture and included me in all of their traditional activities. Things seemed to flow so well that I forgot that I had promised the Arab shopkeepers that I would be back to enjoy their hospitality. As night progressed, the Nuba told me that I could sleep on a cot outside one of their homes. Even though it would have been much more comfortable to stay with the shopkeepers, I was caught up with the experience at hand and decided to stay with the Nuba. It seemed too awkward to run back to the settlement and refuse their offer.

After three days, I realized that I needed to get back to Khartoum, at which time I returned to the settlement to apologize to the shopkeepers and wait for the next lorry going to Kadugli.

Back in Khartoum my stay there was winding down before we were to leave. Mary and I wanted another child, but the doctors in the States said it would need to be by cesarean birth since the twins had been born cesarean. Once a cesarean, a cesarean was always the rule. However, we heard that was not necessarily true and depended on the circumstances during the birthing process. After consulting an obstetrician in Khartoum, we were told that he would be alright with a normal delivery if the baby was born there. So, we proceeded. Leila was delivered normally in May, two months before we were to return to the States.

My memories of Sudan are all favorable. We were blessed to have been there at a time of peace. Although there were a few governmental disturbances, the general population was hardly impacted. Under the recently deposed dictator, Omar Bashir (2019), Sudan experienced severe violence and genocide, and now hopes to proceed toward democracy which now has failed again. Without dictatorial rule, different factions begin expressing their agendas that frequently promote civil war. This sometimes results in separation into separate countries. From the time of colonial emancipation, until the time I was living there, this separation had not happened anywhere in Africa. Since then, Eritrea has become its own country. Independent South Sudan has its trials. Darfur in the far west would like to have its own country, the Nuba would like to break away and possibly join South Sudan. It saddens me to see a country that filled me with such excellent experiences succumb to violence and tragedy.

21. URBAN FARMING IN INDIANAPOLIS

Urban Farming in Indianapolis by Rick Bein, Bhuwan Thapa, Jeff Wilson

Urban agriculture is the practice of cultivating vegetables in or around urban areas. Besides horticultural products, urban agriculture involves animal husbandry, beekeeping, aquaculture, and agroforestry. Increasing urbanization coupled with the challenges of sustainable development in cities have contributed to a growing interest among urban agriculture researchers, policy makers, and stakeholders (Siegner et al., 2018). Growing prevalence of urban agriculture represents a resurgence of vegetable growing where there is a popular dimension of contributing to a greater cause. The chicness of gardening provides opportunities to serve preferences for local produce. Urban Agriculture is also being explored as a strategy to address issues of food insecurity, income inequality and health disparities. Gardening is considered an option to remedy the urban food deserts where there are no grocery stores within a one-mile radius. The myriad potential health & social benefits while maintaining a watch out for toxic air and legacy contaminants in soils.

History of urban agriculture in Indianapolis

Urban agriculture has been a part of the landscape of Indianapolis begun in the 1840s (Jo Ellen Meyers Sharp, 2019). German immigrants brought with them their greenhouse technology and developed numerous urban farms on the south side of the city. From that time, they provided a significant supply of the city's vegetable needs until the 1970's. Year-round production was made possible by heating the greenhouses in the winter using mostly coal as fuel. Diminishing evidence of these old greenhouses (Figure 1) can still be found on the south side of the city, in particular along Bluff Road, W. Washington and Rockville Rd. This area was once a hub of vegetable production in Indianapolis. The Indiana Historical Bureau placed an official state historical marker honoring the early German gardeners in Bluff Road Park at Bluff Rd & Hanna Ave. (Sharp 2019)

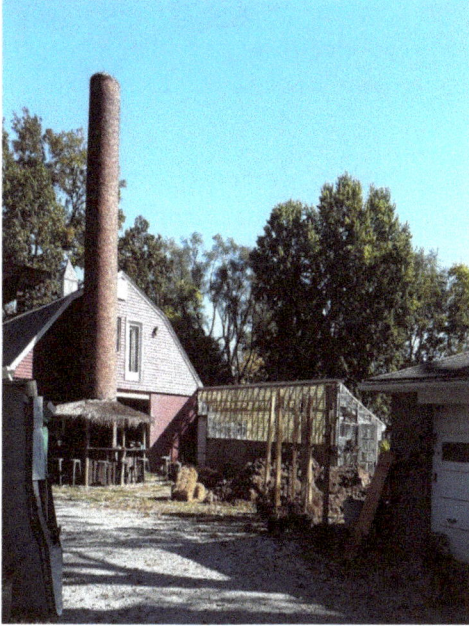

Figure 1. Remains of a 1950's greenhouse and smokestack of a coal fired furnace. Photo by Rick Bein 2020.

With construction of the interstate highway system in the 1960's, many changes to the American landscapes occurred. The Census of Agriculture shows a drastic decline of farmland in Marion County from 58 percent of the county to 23 percent in 1978 and 2.9 percent in 2012 (1). Much of the decline in local vegetable production in Indianapolis diminished drastically with the growth of the interstate highway system enabling refrigerated trucks to bring fresh vegetables in two days from frost free areas in California, Florida, and Texas. As transportation and refrigerated shipping became more efficient, the cost of produce imported from out of state became less than those of the Indianapolis producers who had to bear the cost of heating greenhouses. Gradually, the local vegetable production in Indianapolis was reduced to a few specialty vegetables like herbs and spices.

Indianapolis has experienced a resurgence in the last 10 years for commercial, community, not for profit and institutional gardens (United States Census Bureau). This resurgence in Indianapolis is due to several factors. New low-cost technologies of wind and solar energy and high-capacity batteries has made it possible to provide the necessary heating of greenhouses and plastic hoop towers. With electricity becoming cheaper, greenhouses are more efficient, and an expansion of garden technology has led to increasing local production.

The dormant green thumb is emerging within the county. Community gardening in neighborhoods when vacant land becomes available, and groups of local residents who combine their efforts to begin raising vegetables. Most of these are "home consumption" arrangements that provide food for families. Institutional gardens are sponsored by churches, schools, hospitals and city government and donate produce to food pantries or donate directly to needy families. This aspect has been brought about by Indianapolis City government's support of not-for-profit gardening with land and "startup grants." Grants from businesses and industrial groups have supported the continuation not-for-profit and community gardens. The Purdue agricultural extension program has made a major encouragement effort by educating growers, proving technical assistance and gardening supplies. Other attributes for gardening include environmental conservation, minor reduction of the urban heat island effect, educational purposes at schools, social cohesion of neighborhoods, biodiversity enhancement when combined with bee keeping, and soil health enhancement with earthworm compost.

The Geography Department at Indiana University Purdue University at Indianapolis (IUPUI) launched a study of the spatial aspects of urban vegetable gardening in Marion County with the support of the Purdue Agriculture Extension Service and the Indianapolis Department of Community Nutrition and Food Policy program.

Two student assistants funded by a grant from the IUPUI Center for Service and Learning helped to conduct interviews and organize the responses. AmeriCorps volunteer, Skylar Roeff, assigned to the Indianapolis Department of Community Nutrition and Food Policy contributed greatly to completing a master list started by the Marion County Office of Sustainability. The list was further expanded by contacting other partner agencies like Purdue Extension, Indy Food Council, and browsing through various websites. About 100 gardens were contacted for interviews (Figure 2). On site and telephone interviews were conducted with the gardeners. The data was made available to the City of Indianapolis for planning purposes.

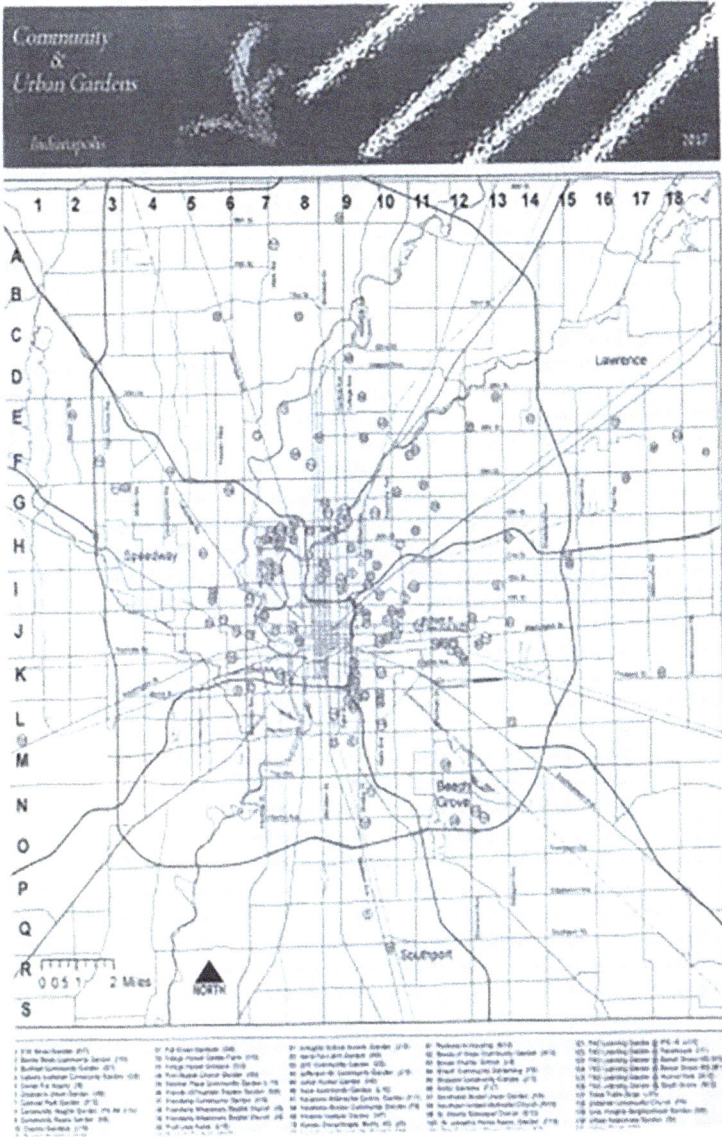

Figure 2: Inventory map of Urban Gardens was developed by the Indianapolis Office of Sustainability 2019.

The survey provided a variety of garden types detailing the purpose for individual gardens, size, type of vegetables grown, and who consumes the produce. Here, differences are highlighted with examples. Private home gardens, although quite prevalent, were not included in this study.

Neighborhood community gardens occur where land is available in back yards or empty lots. They are frequently located in food deserts where there are no grocery stores within a mile. Most neighborhood gardens are organized by one individual who invites others in the vicinity to participate. Raised garden beds are allocated to interested families who may pay a token ($12) to manage and grow what they want for their own consumption. Figure 3 shows James Whitfield (far right) who maintains a neighborhood garden with raised beds shown behind on the ground. Located in a food desert on the near North side of Indianapolis, people generally grow tomatoes, collards, lettuce, and legumes. Serving the people of the neighborhood is the main mission. The major challenge for success depends on one person to carry out the organization, registration, and recruitment of participants. If the lead gardener leaves, there is no one to take over. Water during dry spells is also a challenge. Expensive household tap water is frequently used for irrigation.

Figure 3. Service-Learning assistant, Sneh Shah (middle) interviewing community garden leader, James Whitfield, and his neighbor. Photo by Rick Bein 2018.

Registered Community Gardens are organized with a network of gardeners who share information, conduct workshops and present seminars for anyone who wants to participate. Workshops include food preparation and diet balancing as well as gardening techniques. They also lease out raised beds to landless urbanites. Mixed cropping enables a wide variety of vegetables to be produced. The workshops provide information about pest control, mulching of soil, and companion planting.

Sharrona Moore (Figure 4) found that planting garlic between the broccoli plants kept pests like rabbits, deer and some insects away. She also gives seminars on food preparation and conducts tours. **Lawrence Community Gardens** is a 7.6-acre garden located on 46th Street, just east of Post Road. They partner with Monarch Beverage, located at 9347 Pendleton Pike, and the City of Lawrence to grow produce that is donated to the neighborhood Cupboard and Sharing Place food pantries. Monarch Beverage allows their unused land to grow these vegetables. The missions to promote community and environmental sustainability align perfectly.

Figure 4. Sharrona Moore, Lawrence Community Garden. Photo by Rick Bein 2019.

Figure 5: Carina McDowell, Sugar Grove community Gardeners. Photo by Angela Campbell

Carina McDowell has a good set up with lots of options on-site. They have a tool shed and a makeshift seating area. The mission is to expand food gardening in the neighborhood. Carina is exasperated by the lack of the participation from the community and the discontinuation of funding from the city. She said the Mayor's Action Committee helped with the first year, but nothing after. Challenges are the lack of participation, lack of parking, pests like rabbits and deer and an abundance of mosquitoes. They have a large variety of crops and are exploring ways to cut costs and be more environmentally aware.

Church sponsored community gardens are managed by interested members of the congregations to provide vegetables to needy families.

Figure 6: St. Joan of Arc Garden, Winter season. Photo by Rick Bein 2019

Behind St. Joan of Arc Church A garden with raised beds waits for spring planting. A member of the church oversees the operation and recruits parishioners to participate. Challenges include a shortage of labor, weeds and expensive city tap water.

Figure 7: Burmese Gardens at the 1st Baptist Church. Crops are mixed randomly amongst each other. Note the fence made of fallen sticks from the nearby woods and the plastic bags to frighten away birds. Photo by Rick Bein 2019.

On the property of the First Baptist Church twenty families of Kareni People refugees grow vegetables much in the way they are grown in Burma. Their Garden is primarily for home consumption, and it provides an activity for their senior citizens of the community with productive activity. The Burmese Garden is quite unique as the gardeners use local materials from the wooded areas such as vines and fallen branches to fashion fences. They also practice multicultural cropping by planting different vegetables side by side allowing ecological relationships where plants support each other in various ways. This includes exotic vegetables only known in Southeast Asia. Expensive tap water is provided by the church.

Gardens at schools are quite common which add hands-on activity learning to the curriculum as well as education about aspects of growing, maintaining, and harvesting plants. Many schools in Indianapolis have some form of a garden, only for educational purposes where the children are involved with the gardening processes. These range from a single "raised bed" on a playground to a fully operational greenhouse. At Orchard Elementary School, Vicki Prusinski directs student gardening activities throughout the year in the greenhouse and in a plastic hoop tower. Maple trees on one acre are producing sap (Figure 10) that the children collect and learn how to make maple syrup. The big challenge is scheduling class activities around the seasons and the weather. There are no students to participate during the summer growing season.

Figure 8: Buckets are place low enough so Orchard Elementary students can collect the sap and learn to make maple syrup. Photo by Rick Bein, 2019.

Some hospitals maintain community gardens for their recovering patients who participate in the planting and weeding and in turn benefit from learning food preparation. The Eskinazi Hospital Corporation includes gardens at their neighborhood clinics (Figure 11). Besides health care, their mission is multifaceted, education being first while community interaction and community cohesiveness come second. Many of their patients live close enough that they continue to return to work in the garden after their hospital stay. This contributes to substantial neighborhood camaraderie as well as a healthier diet.

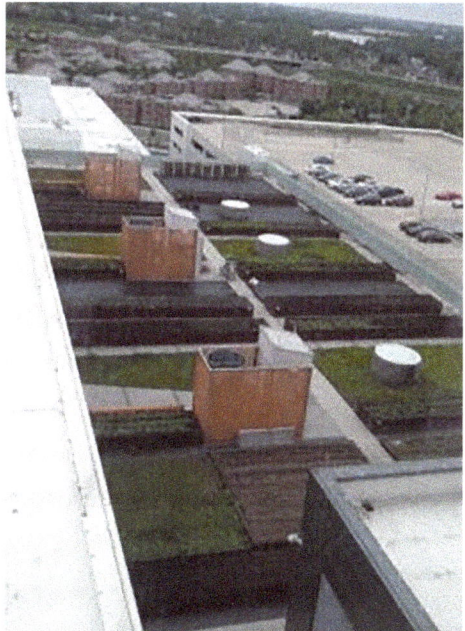

Figure 9: Service-learning assistant Angela Campbell interviews the hospital "green" garden coach at the Brehob Community Center. Figure 10: Sky Garden on the roof of Eskinazi Hospital over looks downtown Indianapolis and Campus Housing.   Photos by Rick Bein.

Eskinazi Sky Farm located on the roof of the Eskinazi Hospital is supported by the Eskinazi Grant Foundation. It has become part of the healing environment of the Eskinazi Hospital on downtown campus. The Sky Farm has 5,000 square feet of growing space dedicated to 25 – 30 different crops and flowers. The initial planting began in the spring of 2014, just after the Sidney & Lois Eskenazi Hospital opened. Rachel White directs all the garden activities.

Education is the main mission providing classes for university dietician program, hospital patients and students. Group tours include elementary school classes. Volunteer groups, mostly students, also enjoy the fresh food. Hospital employees enjoy coffee breaks and the opportunity to meditate. Sky Farm is open to the public from February to November, 24-7. Occasionally it is closed for private events. To access The Sky Farm, take the Green Elevators in the Sidney & Lois Eskenazi Hospital to the seventh floor.

Gardening is done on raised beds using composting with plant material. Plantings include 60 varieties of vegetables but mainly Kale, Collards, tomatoes, and Carrots. "Sedum" is used as a ground cover during the off season while the plots wait to be cultivated. Black plastic cover surrounds the plants for weed control. Worm castings, compost from coffee grounds, fish emulsions and pulverized eggshells provide fertilization. Neem oil is used for pest control. A beehive provides pollination. The main challenge is the expense of water. Since its beginning the harvest has increased every year.. In 2019 more than 3,000 pounds of produce was harvested. Projections for 2020 indicate there will be 3,700 pounds of produce.

Local **government organizations** are sponsoring gardens to support their food pantries. Indianapolis Park Service supports several not-for-profit garden projects around the city. Marion County Soil and Water Conservation Service uses the Indy Urban-Acres Garden (Figure 11) as a demonstration site for proper soil and water management. Indy Urban Acres' main mission is to donate vegetables to needy families. Grocery sized bags are being sent to over 300 families at Senior Citizen communities, Eskinazi Clinics, Glick owned apartment complexes and the Old Bethel Food Pantry. The Garden director Tyler Gough has acquired numerous grants mainly from the Glick Family Foundation, and from the Indianapolis Pacers and Eskinazi Grant Foundation.   Indy Urban Acres cultivates 8 acres producing over 30,000 Pounds of vegetables per year. About 40 different vegetables are grown.

The Indianapolis Parks has set aside a five-acre site on West Tibbs Street (Figure 12) where residents can rent a ¼ acre garden area. The main mission is to provide food for home for consumption. A tractor operator contracts with each gardener (farmer) to prepare the soil for planting. The "Mayor's Garden" has one challenge and that is to provide water for irrigation during dry spells.

Figure 12: The Mayor's Garden" supervised by Indianapolis Parks provides nearby residents with quarter acres to produce vegetables for home consumption. A custom tractor operator prepares soil. The city charges the gardeners $20 per year for five acres. Photos by Rick Bein 2019.

A few "for-profit-gardens" operate in the county. Most of them compete with gardeners from neighboring counties to sell their vegetables at the "Farmers Markets" in Indianapolis. Many of these gardeners have developed online marketing strategies and deliveries. Frequently purchases are made over the telephone or online and the food is picked up by the consumers at the farmers' markets.

Danny Garcia of Garcia's gardens, (Figure 13) surrounds his house, backyard, and a next-door lot with a total 1.2 acres of vegetables. A green house, several hoop structures, numerous raised beds, open soils plots and a compost pile are included. He supports himself completely by selling mostly at farmers markets. He takes online orders and by phone for pick up and deliver to buyers.

Figure 13: Danny Garcia, owner of Garcia's gardens, manages one acre that surround his home with raised beds, greenhouses, and plastic hoop gardens to raise vegetables for sale at numerous farmers markets. His Farmers' market booth is located at the Broad Ripple farmer's market. Photo by Rick Bein 2019.

Founded in 2016, "CRG Grow" is the Cunningham Restaurant Group's own greenhouse. The **hydroponic greenhouse** provides the opportunity to control produce options and quality. Safe growing practices, organic pest treatments, no harsh chemicals, and purified water ensure the produce provided is sustainable,, safe and fresh for the CRG resteraunts. The quarter-acre green house in downtown Indianapolis supplies vegetables to seven restaurants. This electronic facility (Figure 14) includes climate control and water distribution on demand. Organic Peat soil provides a solid growing base. The produce grown at CRG Grow includes basil, mint, thyme, various micro-greens, tomatoes, and peppers. Prior to the Covid 19 pandemic, four employees were required to coordinate supplying vegetables to the scattered restaurants.

Figure 15: Cunningham Restaurant Group's massive greenhouse operates in a downtown Indianapolis building using solar energy for lighting, climate control and irrigation. Diners in 7 mid-western restaurants enjoy the vegetables. Photos by Rick Bein, 2018

Why hasn't large scale vegetable production caught on with the large commercial farms in Indiana? These farms already have a niche in the national economy of industrial agriculture that grow a high tech genetically modified crops. Corn is made into glucose and soybeans are exported overseas. This allows the urban garden phenomenon to fill the empty niche and flourish.

This study looks at the variations in the services that gardens provide according to the institutional categories. These services are measured in terms of production, educational services, and social interactions. The institutional categories include for-profit farms, non-profit farms, educational gardens, and community gardens. This study offers much "food" for thought.

22. FOUR STOREY AGRICULTURE ALONG THE MOZAMBICAN COAST.

Four Storey Agriculture along the Mozambican Coast

F.L. (Rick) Bein and Christopher Hill

Abstract

A system of agro-forestry prevails on the sandy soils on ancient raised dunes in coastal Inhambane, Mozambique. We have given the name "Four Storey Agriculture" to this system as it captures the essential features involving different vertical levels of cultivation. Because of the low fertility of the ancient sand dunes the Portuguese colonial masters allowed most of this land to remain with the local population who continued to make it productive. No chemicals can be afforded by the subsistence farmers who must rely on the natural ecological controls. High biodiversity of crops and native species have reduced fertility loss, risk of erosion and pest and disease problems which would have been encouraged by systems of monoculture. Multilevel poly-culture includes over thirty different crops together with local tree species that synergistically interact to support one another to produce a sustainable chemical and fossil fuel free system. The population density in 2000 was over 67 persons per square kilometer.

Inhambane Province and surrounding districts of Mozambique. Map from National Archives of Mozambique 2006

Introduction

"Four storey agriculture" is the label for the system of agro-forestry that dominates the coastal area of Inhambane Province of Mozambique This system of land use involves the cultivation or preservation of many useful plants growing to different heights (Figure 1). This diverse mix of at least thirty crops includes plants with different life cycles, and those that grow in different environmental micro-zones. Total crop production is increased by dispersing the energy of the strong tropical sun vertically as well as horizontally. A mixture of crops and trees are grown in small family holdings, called machambas, of less than 3 hectares. Coconut trees occupy the top (fourth level) of the agricultural canopy. Fruit trees, including citrus and native occupy the third level. Papaya and bananas are inserted along with the maize to make up the second level. The first level (on the ground) contains vegetables such as beans, tomatoes, greens, sweet potatoes and many more. Since there is no cold season the crops grow throughout the year. Harvesting can occur at any time providing food when needed.

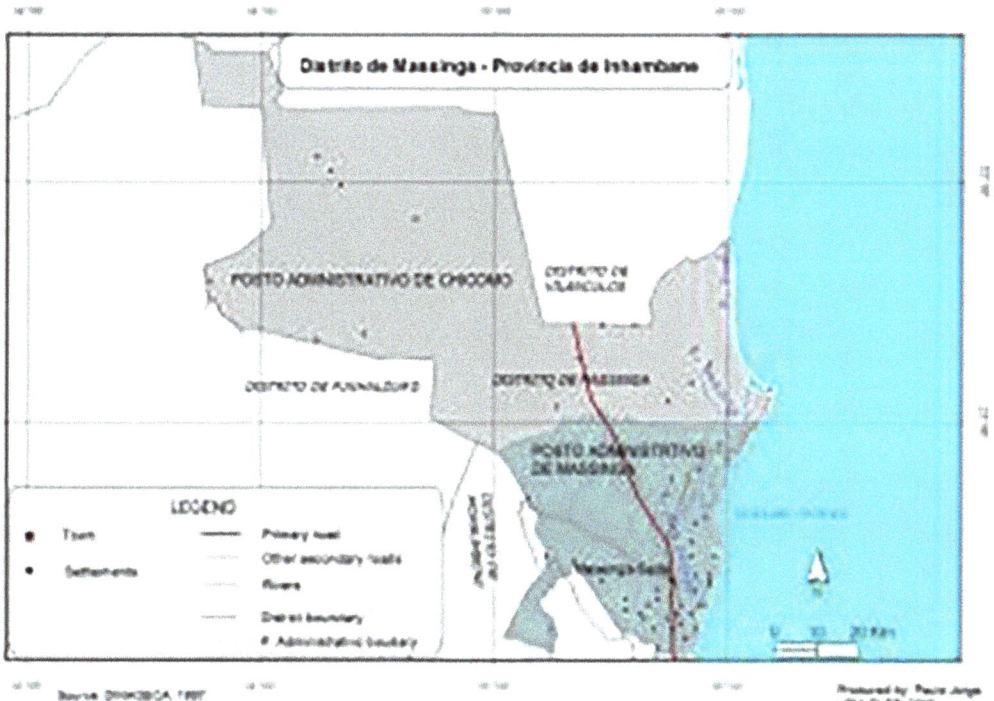

Figure 1. Massinga Administrative Post in the District of Massinga, Province of Inhambane, Mozambique

Massinga, just north of the Topic of Capricorn, is in the transition zone between a tropical and subtropical climate with high rainfall influenced by the warm Argulhas Current along the coast. Although rain can occur throughout the year, there is a main hot rainy season from October to March and another lighter, cool rainy season during July and August. Rainfall is heaviest along the coast yielding over 800 mm (36 inches) per year but diminishes rapidly inland (Figure 2). Average temperature ranges from 19 to 32 degrees Celsius

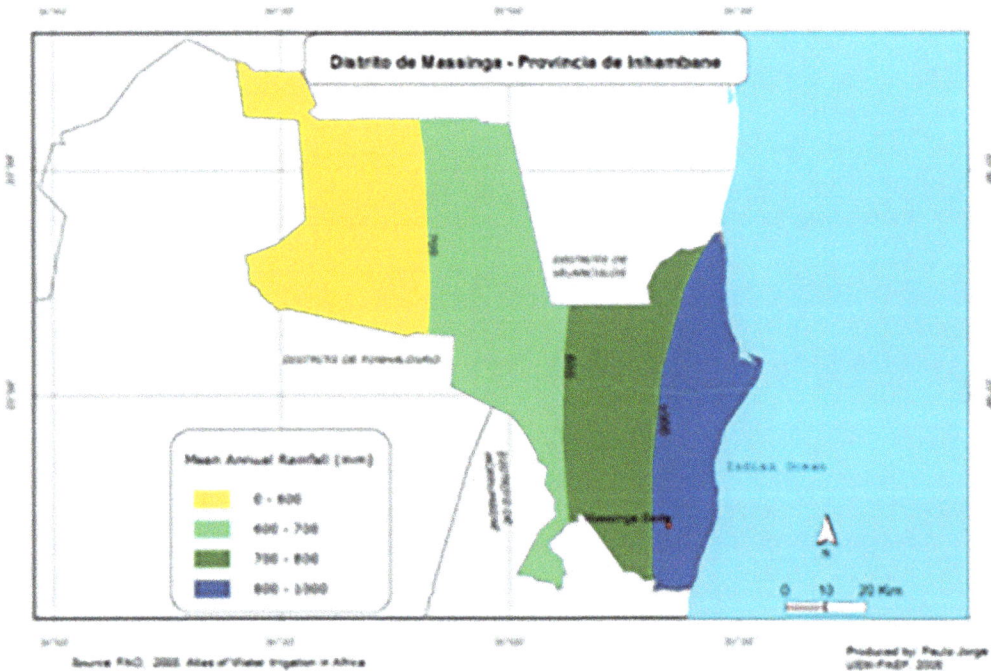

Figure 2. Coastal Inhambane receives adequate rainfall to support agriculture.

Because of the high rainfall the tropical forest zone extends throughout southern coastal Mozambique. The Bantu peoples migrating southwards entered this area nearly 2000 years ago brought with them iron working, livestock rearing and sedentary agriculture. The production of charcoal for iron production and the clearing of land for agriculture have dramatically modified the forest. The agro-forestry ecosystem that remains is the result of major human modifications, but has also retained useful trees from the original forests.

Various crops appear at different heights in this agro-forestry ecosystem. The system has many layers, but can be reduced to four dominant ones as described below:

The top level or the fourth storey is occupied by well-spaced coconut palms. These palms when fully grown generally protrude above any other trees. They spread from the coast covering all the area except for permanent wetlands. Most of the coconut palms are residuals from plantations developed during the colonial period.

The third storey includes introduced domestic trees such as cashew, citrus and mango combined with many native wild trees and bushes. Generally the native species are useful species providing the population with many different products including fruit, material for construction or artifacts, fiber, medicines and firewood. This layer ranges in height with some trees, such as the Mango and the indigenous waterberry, almost as high as the coconuts to others such as the citrus, and guava that only reach 3 meters.

The second storey is occupied by short lived crops that grow upright off the ground and includes cassava, corn and sorghum.

The first storey or ground level and includes peanuts, cow-peas, pumpkins, wild invasive ground cover such as cacana (*Cucurbitaceous*) *Momordica balsamina* and a variety of vegetables. The local people have developed ways to add the cacana to their meals.

We compare the four story agriculture with the natural forest, a multi-storied natural environment. Our designation is keyed on the presence of the coconuts in this area one of the two main traditional zones of coconut palms in Mozambique (**Figure 3).**

Figure 3: Four storey agriculture starts with the coconut palms that can be seen towering over the fruit trees, which in turn partially shade the cassava plants. The emerging cowpeas capture the remaining sun light that reaches the ground. Many other plants are interspersed in small quantities including those serving medicinal and ornamental purposes.

Crop cycles

Plant life cycles and environmental micro-zones also add agricultural diversity to this high precipitation coastal area. Plant life cycles vary as crops mature through out the year. Some crops like cassava **(Figure 4)**, papaya, bananas and coconuts can be harvested at any time during the year while the annual crops and tree fruits must be collected at the end of specific seasons.

Figure 4: Papaya, bananas and coconuts have no season can be harvested at any time during the year.

The life span of the different crops (Table 1) offers another element of diversity as some like the coconut live for decades while others only a few months. With crops grown in level one, rotations occur starting with plowing to start a two or three year cycle where maize, legumes and vegetables are grown together with pineapple and cassava. Once the maize has matured its leaves are harvested to allow more sunlight to reach the lower and slower growing plants. The cobs may be left on the maize stalks until they are needed for food. The lower growing plants many of which have been climbing on the maize stalks, now surge upward to claim the sunlight. In time those are harvested allowing opportunity for the slower growing crops of cassava and sugar cane to take over the canopy.

Cycle Time	Crops
Long	Palms and Fruit Trees
Medium	Bananas and Papaya
One – two year	Cassava, Pineapple, Sugar Cane
Annual	Maize, Cow-pea, Peanuts, Vegetables

Table 1. Crop life cycles the four storey agriculture Inhambane, Mozambique

Physical environmental zones

The physical environment of the Massinga coastal area offers three different micro-environments. These three distinct habitats occupy the immediate study area: the beach, deep ancient sand dunes that extend for many kilometers inland and the wetlands that emerge in low areas between the ancient dunes. These topographic habitats are described in the following outline.

Beach

The beach is the active natural system dynamically controlled by waves, currents, storms and winds. In terms of agricultural land use it is limited to few crops cultivated on recently formed sand dunes adjacent to the ancient dunes. For practical purposes this zone is not considered part of the study.

Ancient sand dunes

The ancient sand dunes were formed during prior geologic periods when the ocean encroached further inland. It is also possible that the gradual uplift of the African Plateau raised the coastal formations and stranded the dunes well above the present ocean level. Much of Mozambique resides on this coastal platform of elevated sand. With respect to land use it is on these sands exceeding 100 meters in depth where most of the four storey agriculture takes place.

Wetlands

Wetlands include streams, lakes and low lying poorly drained areas between the dunes. The better drained parts of these humid lands, though small in area, are highly prized for their year around moisture content that support water needy crops.

Environmental Micro-Zones

In the above description of the physical environmental zones, we find that the ancient sand dunes and the partially drained wetlands are incorporated into four storey agriculture. The machamba (where the most intensive short term crops are grown) can be located on either of these environments. But of equal importance is the area around the homestead (called a quintal[2]). These three micro environmental zones contribute to the system, each containing its own diversity and variety of crops. See Table 2. Most of the machambas are located on ancient well drained sand that have limited fertility but sustains a diverse array of crops dominated by coconuts mixed with other crops and some other trees. While in between the ancient sand dunes are small low lying humid areas where water accumulates and supports a somewhat different array of crops (Figure 5). Here bananas, sugar cane and horticulture thrive.

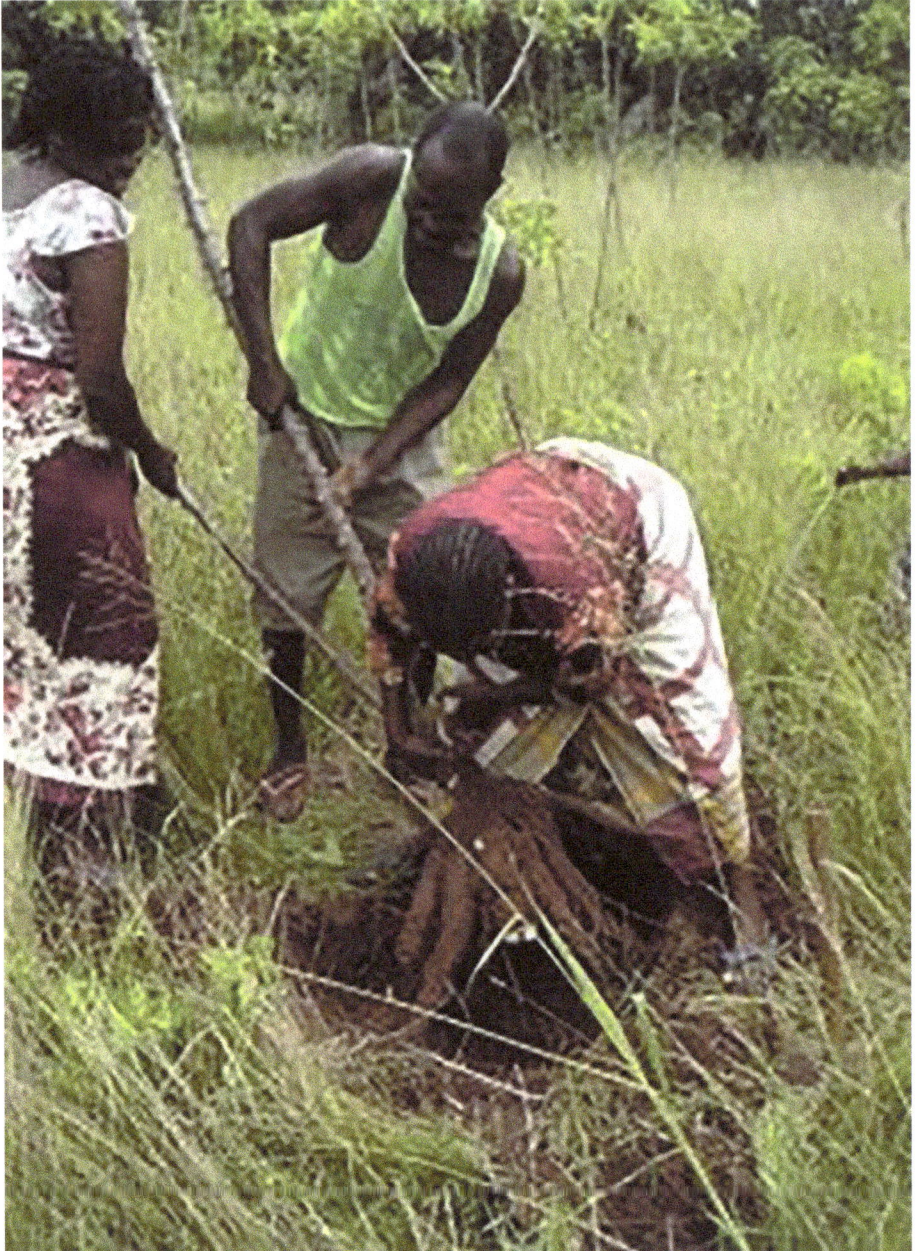

Figure 5: Humid machamba showing the harvest of cassava.

The dispersed rural settlement pattern that prevails in coastal Inhambane Province fosters the quintal micro environment that allows for a highly concentrated mix of crops. Around the quintal many different useful plants are cultivated in small quantities **(Figure 6)**. Here spices, ornamental and medicinal plants are grown for immediate use. Also, this is where the farmer will grow the more recent plant acquisitions with which she can maintain regular contact. Because of the proximity to the dwellings, it is easier to attend to the needs of plants. Hand irrigation and the application of recyclable household wastes help to sustain these more delicate plants. Normally the quintal is located on the ancient sand deposits, but contains all the crops grown in both the humid and upland areas.

Environment Type	Upland	Humid	Quintal
Condition	ancient dunes	wetland borders	regular care
Crops:	coconut trees	Bananas	bananas
	fruit trees	fruit trees	fruit trees
	wild fruit trees	wild fruit trees	wild fruit trees
	other useful wild trees		
	cassava	sugar cane	sugar cane
	maize	sweet potato	sweet potato
	Cow-pea	horticulture	horticulture
	peanuts		medicinal plants
	squash		coconut trees
	tomatoes		spices

Table 2. Products of four storey agriculture by micro-environmental zone along the high rainfall coast of Inhambane, Mozambique.

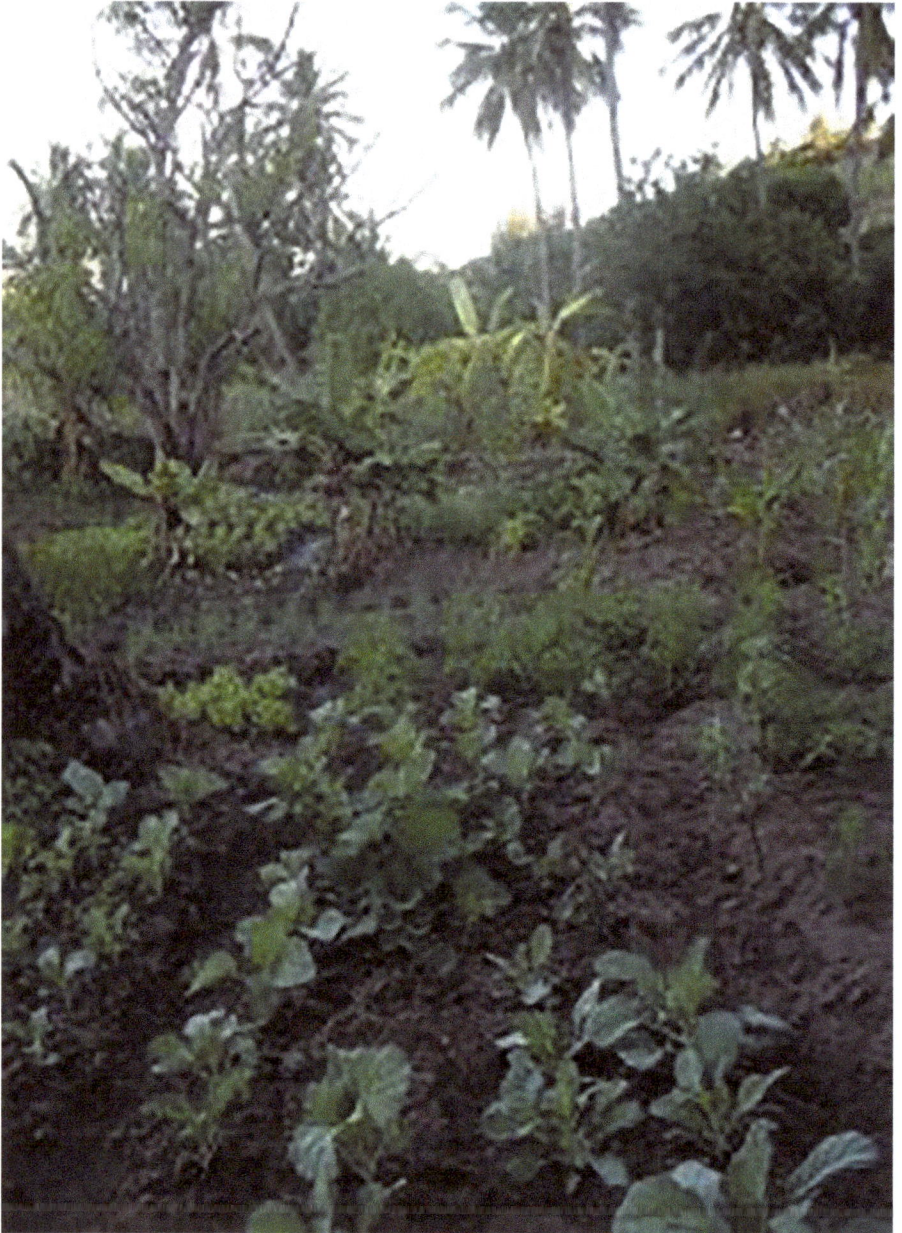

Figure 6: the biodiversity around the quintal is very high because the plants receive regular attention. Here the indigenous Waterberry tree (Syzygium cordatum) provides fruits that are eaten but also made into alcohol.

Tropical Biodiversity

Bio-diversity is greater in the humid tropics because environmental limitations are less than in the temperate regions. Winter kill does not exist and prolonged drought is infrequent. As a result the natural biodiversity of pests like weeds, insects and diseases are much more prevalent and are extremely difficult and expensive to eliminate in the tropics. However, the natural predators of the weeds, pests and diseases are also more diverse and react to control their numbers. The higher biodiversity reduces pest and disease problems which would have been invited by monoculture. It is the biodiversity itself that controls (not eliminates) pests. A study done by K.S. Powel[3] in Papua New Guinea in 1998 shows that insect pest species are outnumbered by 50% by their insect predator species. The conditions here are similar in the Massinga District machambas where there is no chemical or manual control of insects and the biodiversity itself helps to keep pest populations in check. Fallow periods in the two lower stories, also help control for pests and diseases while effectively providing space for a few livestock to graze. In general crop biodiversity minimizes the risk of insect infestations and plant diseases while the crop mix does not deplete the weak sandy soils as rapidly as monoculture.

Land use

As stated above the four storey agriculture is typified by coconut trees spaced from 10 -15 meters apart with other useful trees dispersed below. The crops of the two lower levels are rotated between fallow periods. Here controlled burns are normally used to clear the low level under- story vegetation while carefully protecting the existing trees and any new useful wild trees that might have grown during the fallow period. The fallow cycle normally varies from three to fifteen years.

Many wild African trees species grow spontaneously in the machamba and fallow areas and those that have utility are allowed to remain. Most of these are fruit trees, but others are used for fiber, construction or medical purposes. There are over eighty different useful plants growing in the study area. This includes twenty four identified useful wild fruit tree species[4].

The different crops support each other with nutrients. Particularly important are the legumes cow-peas and peanuts and a number of common trees that deposit nitrogen in the nutrient starved soil. Taller crops like trees, cassava and maize support the climbing plants. Farmers seek shady spots to plant sensitive vegetables. Even though the coconut trees take a lot of water out of the soil, their shade and the shade of the fruit trees reduces evaporation and makes more moisture available for the lower growing crops. The ground cover crops reduce the loss of humidity in the soil, reduce the quantity of weeds and reduce soil erosion.

Using air photography provided by "Google Earth"[5] over the internet, the land use can be clearly demarcated when the high resolution imagery is enlarged. Coconuts palms can be counted groves of bananas and other fruit trees demarcated, and full grown cassava plantations recognized. It general area of land use such as annual crops and fallow land can be clearly identified. We have verified that what we see from above is truly what we see on the ground. See **Figure 7**.

Figure 7: Diversity of cropping can be seen in this Google Earth image. The quintal in the center of the circle is surrounded by many types of wild domesticated trees species.

Comparison with shifting cultivation

The four storey agriculture must be clearly differentiated from the practice of slash and burn shifting cultivation. Typically slash and burn tactics indiscriminately destroys the forest to make way for crops. Cropping continues on the same land for 1-3 years followed by forest fallows that may last decades before they are cut and burned again. The sustainable human population in shifting cultivation is very low and clearings are dispersed over wide expanses of forested land. Hunting and gathering also contributes to sustain the farmers. As population pressure increases the forest fallow period becomes shorter and shorter until the cultivated land is greater than the fallow land.[6] New strategies evolve that the farmers find necessary to over come increasing pest problems, depletion of soils and reduction of available land.

Our study area is part of the relative densely populated coastal zone of Mozambique and occupies the coastal part of Massinga Post in Massinga District. In 1997 the population of Massinga post was 138, 871 with an arithmetic density of about 67 persons per square kilometer.[7] The actually population density of the study area is higher that this because much of the administrative post includes undesirable land further from the coast.

In the coastal zone of Inhambane, the four storey agriculture has evolved with the Bantu people whose Iron Age capabilities allowed them to alter the forest. As the population increased, the forest became "at risk" and modifications were made to compensate for the loss of the hunting and gathering that supplemented their lively hoods. A deliberate strategy has developed of selectively maintaining forest plants. Fire became used in a limited way to clear the ground below the useful trees. Essentially this strategy involved limited or controlled burning and the preservation of useful wild plant species with in or between the cultivated areas. Within the memory of the living people, large wild life has disappeared.

Fallow

The fallow land in the four story agriculture is still productive. Fallow refers to land where the lower two stories, the annual and semi-annual crops are not grown. On the fallow, the coconuts and useful trees continue to produce harvests while tethered livestock graze the grasses and shrubs that grow up naturally. Un-harvested cassava plants remain amongst the new bushy growth. "Grass fallow" characterizes the land in the early years but as more herbaceous and woody plants grow it is frequently called "bush fallow". [8]

Land may remain fallow from 2 to 18 years. With each additional year, the vegetation on the fallow area becomes denser and more diverse, eventually becoming impassable by humans. Many wild plants grow to maturity and wild trees join the canopy of coconuts and other useful trees. Climbing plants grow up into the trees seeking a piece of the sunlight. Plants and animals that would not be allowed to exist where the annual crops are cultivated can now flourish. Some of our respondents say that many of fruits that die in the controlled burns will only survive where the fallow period is long. The biodiversity is highest in the older fallow areas and have become a store houses of fruits, medicines and plants used for fiber, glue and construction material. Traditional healers often maintain exclusive areas of medicinal and sacred plants that extend the fallow into many decades **(Figure 8)**

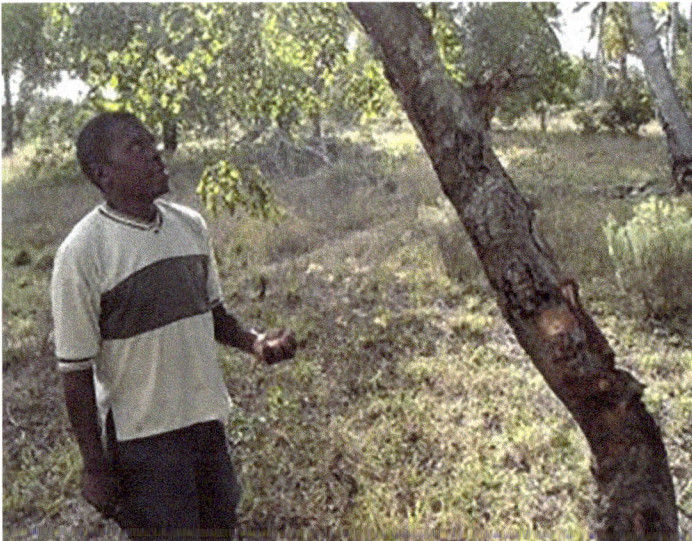

Figure 8: Wild African trees with medicinal value are maintained in the machambas. This local herbalist describes the qualities of this tree.

There seems to be no certain length of time that a piece of land will be allowed to remain fallow. The land closest to the quintal will experience shorter fallows. In some cases an area has been fallow for a long time because its owners have gone away or the heirs have not take the same interests in farming as their parents. In many cases the coconuts go unattended and the neighbors end up harvesting the copra.

Our respondents tell us that there are small patches of long fallow that have emerged with lots of wild fruits and because these have become favorite areas for the children. Many of the wild fruits eaten are produced on smaller trees or vines that grow quickly after the burning and clearing takes place. Likewise when there is a vigorous growth of useful plants in a fallow area the farmers will clear around them and be careful not to burn them when they are preparing machambas. An example of such a wild tree is the Mopani (colophospernum mopane) which has an unusual benefit of attracting mopane worms (imbrasia belina,) which are considered a delicacy by the locals.

Another respondent says that it is a lot of work to clear an area left in fallow for a long time, but it is worth it because the dead leaves and wood that has fallen on the ground acts as fertilizer for new crops. She adds that the accumulated wood is good for building material and for fuel. By allowing the fallow to continue for more years she could benefit from the harvest of this wood.

Respondents say that they often will allow the fallow go longer than 2-3 years when they use it to pasture pigs, cows, goats and donkeys, but only when it is a good distance form the house. The closer fields will be returned to annual crops as soon as possible.

We have observed many of the fallow areas and have noted many differences in appearance. In some of the older fallows the vegetation has grown up to the density described in previous paragraphs while others have become dominated by one species of tree. At one such site we found a large number of Tzondzo (Fabaceae) Brachystegia spiciformis trees, useful for its bark to make string and canoes. At this particular site these trees reached their full growth of 6-8 meters but did not create a closed canopy. Sunlight reached the ground in many places and the growth of other species was scant and stunted. The coconut trees in the area only had few coconuts. It is hard to imagine that this fallow area had not experienced human intervention. Yet there is much to understand about the nature of the Tzondzo tree with regard to its specific biology and chemistry.

Another long fallow site that we visited contained healthy coconuts with a large number of waterberry trees locally known as ticori trees (Myrtaceae Syzygium cordatum, (See fruit in Figure 7 above.) It would be expected that as the fallow progressed the biodiversity would increase. The frequency of the ticori contradicts this. The ticori tree grows wild and survives controlled burning and is encouraged by the farmers because of its delicious fruits and for its high quality of distilled alcohol. It may be that the ticori trees had been allowed to accumulate over successive clearing times and fallow times, so that there dominance in this particular fallow was notable.

One site has been fallow 22 years and contains no coconuts. Its owner had no interest in farming the area. Nearby residents say that the area had been a community horticulture garden before it was abandoned. Today it represents a rare preserve of vegetation that is proceeding into more advanced stages of ecological succession.

The fallow periods are diminishing in the coastal area of Inhambane. This is recognizable by the increasing shortages of construction wood which would be provided in the long fallow areas. There is an increasing demand for construction wood for housing which is now coming from the dry interior of the province.

Environmental and cultural limitations

The four storey agriculture developed from the natural forest and maintained much of its biodiversity. The useful forest plants have been retained, while over an extended period new crops have been introduced. Natural forces tend to fill as many ecological niches as possible, while human intervention tends to encourage only useful species. The principal environmental limitations of biodiversity are the dry season, variability of rainfall, occasional controlled fires, culturally selective elimination of non-useful wild plants, limited animal grazing and cultivation of annual crops.

The infertile soils which might be considered a limitation for monoculture is less of a factor in the polyculture of four storey agriculture. Three soil samples were taken in July 2006 in three locations (Table 3) in the study area.[9] The first was taken in the intensively planted area in the Quintal, the second was taken in the area of bush fallow where the soil had been un-tilled for the longest time of 22 years and the third was taken in area of grass fallow becoming bush fallow. All of the pH levels were slightly acidic with low levels of phosphorus and potassium. Nitrogen although low, was in the normal range expected for sandy soils

Development for survival

This survival strategy system has evolved over the last 2000 years beginning with the immigration of Bantu people who have added to their native crops by including useful exotic plants whenever they became adaptively available. Only the cow peas, sorghum, millet, okra, bambara ground nut, squash, water melon, and yam originate in Africa. Introductions came first from Asia and later from America. All the domesticated fruit trees are introduced species although there a many species of edible wild fruit trees. A few native African fruit tree species are becoming domesticated.

Historically four storey agriculture offered a system that reduced risks so farmers could provide for themselves when crops failed. It helped them survive many years of political instability and an eighteen year long civil war. The diversity of annual crops and fruits distributed food harvesting over a greater period of the year. The diversity of foods offered a self-sufficient nutritious diet. If there was a drought, some crops survived better than others. See Table 4. The system offered a conservative strategy that has served them time and time again.

Coconuts (cocos nucifera) are the key cash crop, but a wide array of other crops are also sold. Alcohol distilled from fermented fruits and sugar cane is also a major source of income. Most crops are grown for home consumption. Coconut products also a major item of home consumption. Coconut milk is a key component of many dishes consumed by the local population. Other crops, like peanuts may be grown for both sale and home consumption particularly when there is a surplus. See **Table 4**.

Commercial	Own Consumption
copra and coconuts	coconut products
alcohol distilled from fruit	maize
Peanuts	peanuts
artisan products	Cow peas
citrus, mango, guava, papaya fruits	fruits (cultivated and wild)
firewood and charcoal	cassava
construction material	sorghum
livestock	products of the Natal Mahogany
cashew nuts	sweet potatoes
meat	meat
	vegetables
	mopane worms

Table 4. The Products of Four Storey Agriculture, Massinga, Mozambique

The Coconut Zone

Coconuts are traditionally grown in the coastal zone of Inhambane. The early Portuguese arriving in 16 century reported the presence of the coconut trade at the city of Inhambane[10]. . The relative high rainfall and the tolerance to the sandy soils made it an ideal habitat for the coconut palm. More coconut plantations were encouraged for the export of copra. Today Inhambane contains the largest percentage of farms with coconuts than any other province in Mozambique[11]. See **Figure 9** The area of coconuts is currently spreading inland and southward.

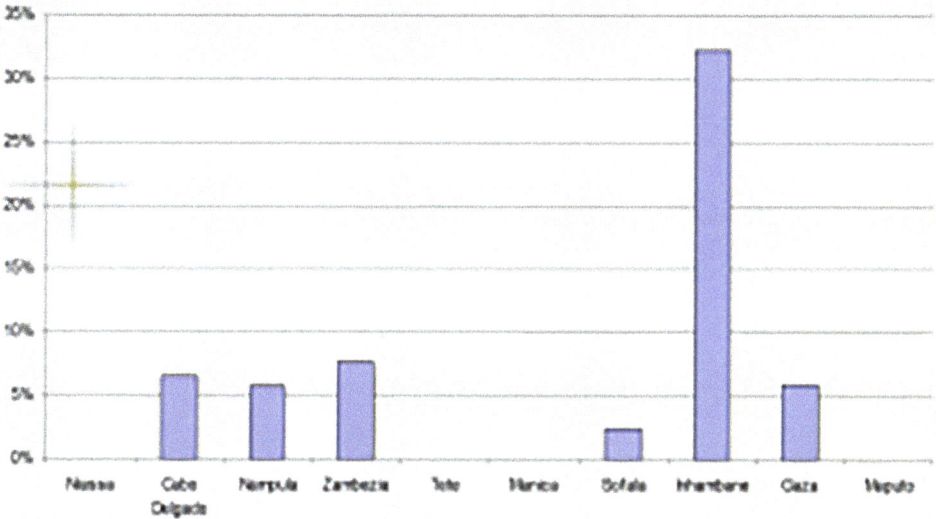

Figure 11. Percentage of Farms with Four Storey Agriculture in Moçambique.
Source: Censo Agro Pecuário do Mozambique 2002

Figure 9: Percentage of Farms with Four Story Agriculture in Mozambique
Source: Censo Agro-pecuário. 2000. Jointly published by Ministério da
Agricultura and Instituto Nacional de Estatistica, Moçambique.[12]

There has recently been a revival of the copra trade, but this trade still only buys a small share of the total farm production. Whatever the case, the local population continued using the coconuts and the palms in their traditional manner as they provided a number of other products that the community needed. See Table 5.

Coconut	Other parts of the tree
copra	palm wine
cooking oil and soap	wood

coconut milk	fronds for roof and walls
coconut water	wall braces
ladles	brooms
husks for:	water container for still
cushions	pins
fuel	
rope	

Table 5. Coconut tree uses in Massinga, Inhambane, **Mozambique.**

Citrus and mangos has been a long time partner in the four storey system. Tangerines were added in the 1930's.[13] The copra commerce declined seriously during the civil war but has experienced some recovery recently.

There are two major coconut zones in Mozambique, Zambezi and Inhambane. The patterns of cultivation of these two zones are quite different. Zambezi Province consists mainly of large scale plantations of improved varieties, whereas cultivation in Inhambane is smallholder agriculture growing traditional varieties. Today there is a major problem with the improved varieties as huge areas of palms are dying of disease.

Use of "Google Earth" for Analyzing Land Use

The use Google Earth has made it possible to see houses and trees in great detail. The detail is sufficient to be able to clearly identify coconut palms. Even where there is dense tree cover palms can be seen clearly when they are tall enough to stand out above the other trees. It has also been possible to clearly locate other areas of vegetation including areas of annual cultivation even when this is under partial tree cover. It is also possible to clearly identify areas of long fallow where there is dense scrub vegetation under tall trees.

A study area of 80 square kilometers has been identified. This area is bounded on the east by the Indian Ocean, on the south be a red sand road, on the west by the national highway and a minor highway leading to the village of Rio das Pedras and on the north by the wet margins and swamp of the river Rio das Pedras. The interest is to use analytic software to attempt to make an analysis of the entire vegetation cover of the study area and also identify the extent to which coconut palms are growing west of the area.

The following classification of land use can be identified

River and swamp lands

1. Cultivated areas with all year round humid soils adjoining rivers and swamps with little tree cover generally containing either 1.) bananas or 2.) sugar cane

2. Homesteads clear of vegetation often with a surround of dense trees

3. Cultivated fields normally with some light tree cover

4. Bush Fallow – Dense areas of trees and scrub vegetation

5. Areas that are under short term fallow

The coconuts are clearly identifiable by shape and also appear to have a different color from other trees. It is therefore also possible to distinguish areas in which most or all the trees are coconut palms and areas where the coconut palms are mixed with other trees.

Determining the Limits of the area of Coconut Palms

The study area is for the most part an area of coconut palms apart from the wet zones. West of this area the density of coconut palms declines. Whereas, in the study area coconuts are generally spread throughout the countryside but to the west coconut palms tend to be found in distinct clusters associated with individual holdings. As coconuts are not naturally dispersed on land their presence indicates deliberate planting, so they tend to be grown in organized rows or clusters.

1. Concentrated coconut palm zone
2. Intermediate zone where coconuts palms are found in reasonable quantities normally in clusters
3. Rare zone where there are few palms
4. Non coconut zone.

More general estimate of coconut palms

The study area is an area of concentrated coconut palms. Coconuts are found along the coastal area throughout the province of Inhambane. The concentrated zone occurs from Massinga district south through most of the other coastal districts of the province. We would like to provide a indication of the complete area of coconut palms with possibly a classification into the concentrated area and the intermediate zone.

The Importance of livestock

Small animals and poultry including goats, pigs and chickens contribute significantly to the diet of the people and are also sold to bring income. However there have been serious disease problems. Newcastle Disease and chicken influenza are annual events but can be preventable by vaccination distribution. Pestiswina Africana periodically almost destroys the entire pig population. There are also diseases that attack the goats. These diseases serious reduce the productivity of the four storey agricultural system.

Traction animals are an important source of power for farming in this system (See Figure 10). Although the numbers of these animals diminished greatly during the long civil war of the 1970s and the 1980s they are recovering their numbers with time. Plowing with oxen and donkeys facilitates the cultivation process with the annual and biannual crops. With out them soil must be tilled by hand hoeing.

Figure 10: Bullocks being used to plow the soil before planting maize.

The Role of Gender in the Four Story Agriculture

In rural Inhambane there is a ratio of 175 women for every 100 men[14]. In most cases the persons responsible for the machambas are women. There has been a long tradition of men going to work in the mines in South Africa. Remittances sent home contributed significantly to the local economy. Children are important assets to this labor intensive farming. When men are part of the farming process they perform the heavy work like land clearing and plowing. They are also much more likely to engage in non agricultural activities to earn extra inc

The authors interviewed both men and women farmers about their roles. Although some activities are gender specific, many activities are not rigidly defined by gender (Figure 11).

Figure 11: Preparing food for drying is typically the job of women and girls.

Women play more active roles in the agriculture process, but men are also involved. Normally a certain task like carrying produce from the machambas is done by women, but men can be seen doing the same task. Table 6 details gender roles.

Gender Role	Women	Men	Girls	Boys
Grating cassava	x		x	
Selecting seed	x			
Clearing land		x		
Planting	x	x		
Alcohol preparation	x			

	Col 1	Col 2	Col 3	Col 4
Plowing		x		
Hoeing	xx	x		
Drawing water	x		xx	x
Collecting wood	x		x	x
Climbing coconuts trees		xx	x	x
Preparing coconut milk	x		x	
Tapping coconut wine		x		
Fishing from boats		x		
Hunting		x		
Trapping small animals		xx	x	
Food preparation	x	x	x	x
Cooking	x		x	
Feeding chickens	x		x	
Feeding pigs	x	x		
Care for cattle		x		x
Care for goats	x	x		x
Preparing Cashews	x			
Collecting Cashews	x	x	x	x
Cutting sugar cane	x	x		
Weaving palm fronds.	x	x		
Weaving mats	x	x		
Building houses	x	x		
Making charcoal		x		
Cutting construction cane	x	x		
Child care	x		x	

Treating Illness	x	x
Weeding	xx	x
Carrying Harvest	x	x

Table 6. Gender roles in Sota, Massinga District, Inhambane,

Mozambique. Adolescents are included as adults rather than children. xx = more likely. Source Maria Rosa Guambe[16

Conservation of Labor in Four Storey Agriculture

In the machambas of coastal area of Massinga there is never a time when there is nothing to do. Because of the diversity of crops and protected useful wild species, many types of activities create a high demand for human labor. Animal traction is used by some families but is the only power source used. Only the tasks that are essential to the farming process are performed leaving little time for the appearance of neatness. Jobs like clearing large dead trees or removing termite mounds are left undone because they have no priority.

Controlled burning is undertaken to clear the land. It saves labor that also helps reduce pests. Dead trees may remain for some years while the small controlled burns eat away at the wood. Pieces of the wood are also collected for fuel when necessary.

The normal settlement pattern of this area is dispersed farms rather than centralized villages. This allows the small holders to live permanently on their farm land minimizing the time taken to reach their machambas. Forgoing the safety of a village, living at the work site drastically reduces travel time and results in significant conservation of labor.

Most of the labor in the four story agriculture is involved with the annual and biannual crops and the most potentially demanding task is weeding. By having good ground cover crops like cow peas reduces the need for weeding. Land preparation can be equally arduous particularly if the land being prepared has been fallow a long time. Here the use of controlled burning reduces labor.

Future of Four Story Agriculture

Four storey agriculture offers a highly sustainable system of continuous agriculture which has been existence for at least one thousand years. The highly diverse cropping system enables the successful cultivation of low nutrient sandy soil. The synergistic mixture of over 30 crops compensates for inadequacies of the soil. The different crops support one another as some add nutrients to the soil that the others need, while others seem to deter pests that prey on the former. Where weeds flourish, the people find a way to incorporate them within the system.

It is also an environmentally sustainable system that appears to reasonably optimize the uses of natural resources. The non use of fossil fuels in four story agriculture makes the system sustainable in light of the world's current fuel crisis. Of course the affordability of fossil fuels is beyond the reach of this rural Mozambican region.

The system operates mostly outside the formal economy and the people prevail in what the modern world would call abject poverty yet there is adequate food for local consumption. The system also coexists with high labor migration of men to the Republic of South Africa. Without this remittance income the population would be much poorer and its sustainability threatened.

There is a great potential to increase the income of this system with improved organization and marketing of fruits. At the moment there is great spoilage of fruits for lack of infrastructure. While the annual occurrence of crop surpluses would make it possible to increase farmer income it would also contribute to the economy of Mozambique if infrastructure such as roads, agro-industries and micro financing were available. The marketing and protection of small livestock and poultry could also contribute to the economic well-being of the community as diseases frequently decimate large numbers of livestock and poultry.

References:

Bein, F. L. (Rick) and Christopher Hill. (2009). Four Storey Agriculture in the District of Massinga, Province of Inhambane Along the Mozambican Coast,

Journal of Geography, vol. 52, no 4.

Berkes, Fikret. (2008). Routlege, New York.

Boserup, Esther. (1965) The conditions of Agricultural Growth: The economics of Agrarian Change under Population Pressure. Chicago: Aldine Press.

Censo Agro-pecuário. 2000. Jointly published by Ministério da Agricultura and Instituto Nacional de Estatística, Moçambique.

Google Earth (2006) Virtual Globe Program.

Newitt, Malyn D.D.(1995) A History of Mozambique. London: Husrt & Comapny .

S. Powel. 1997. Biodiversity Inventory of the Kamiali Wildlife Management Area. Edited by F.L. Bein, Papua New Guinea University of Technology, Lei, PNG.

Renciamento Geral da População e Habitação do Mozamibique. 1997. Instituto National de Estatistica do Moçambiaque.

University of Eduardo Mondlane Levantamento da Fauna Bravia e Avaliaçao das Plantas Uteis em Morrungulo, Distrito de Massinga, Provincia de Inhambane, Unpuplished Report Maputo 2003 supplemented by additional field work by the authors

[1]   Mozambique resides in Southeast Africa along the Indian Ocean extending 2500 km. along the coast from the border of South Africa at 26º 51' south 32º 53' east including borders with Swaziland, South Africa, Zimbabwe, Zambia, Malawi, and Tanzania at 10º 28' south, 40º 26' east. It occupies the coastal plane of Africa abutting to the west with the African Plateau. It is approximately twice the size of California and contains 10 provinces and houses 20 million people.

[2]   Quintal is the Portuguese term for door yard which in this study includes the residence, surrounding buildings and area of specialized crops.

[3]   K. S. Powel. 1997. p 38.

[4]   University of Eduardo Mondlane

[5]   Google Earth 2006.

[6]   Boserup, Ester. pp 75-80.

[7]   Renciamento Geral da População e Habitação do Mozamibique.

[8]   Boserup, Ester. pp 65 -66

[9]   Soil samples taken from the top 50 millimeters included litter layer when present. Analysis was completed by the Soils Laboratory staff at the University Eduardo Mondlane.

[10]   Newitt, Malyn, p 29.

[11]   Censo Agro-pecuário. 2000.

[12]   Ibid

[13]   Manuel Zunguzi (elderly resident farmer) private interview. July 2006

[14]   Christopher Hill. "Report on Gender in Mozambique"2002 Unpublished

[15]   Censo Agro-pecuário. 2000

[16]   Maria Rosa Guambe, Resident farmer interview July 2006

23. FARMING IN ICELAND

Farming in Iceland

Iceland is not commonly thought of as a place where farming can prosper. Because of its far north latitude, just below the Artic circle, one would expect a climate far too cold for agriculture. However, because of the Gulf Stream Current bringing warm water from the region of the Gulf of Mexico, Iceland is warmed enough that vegetation can prosper.

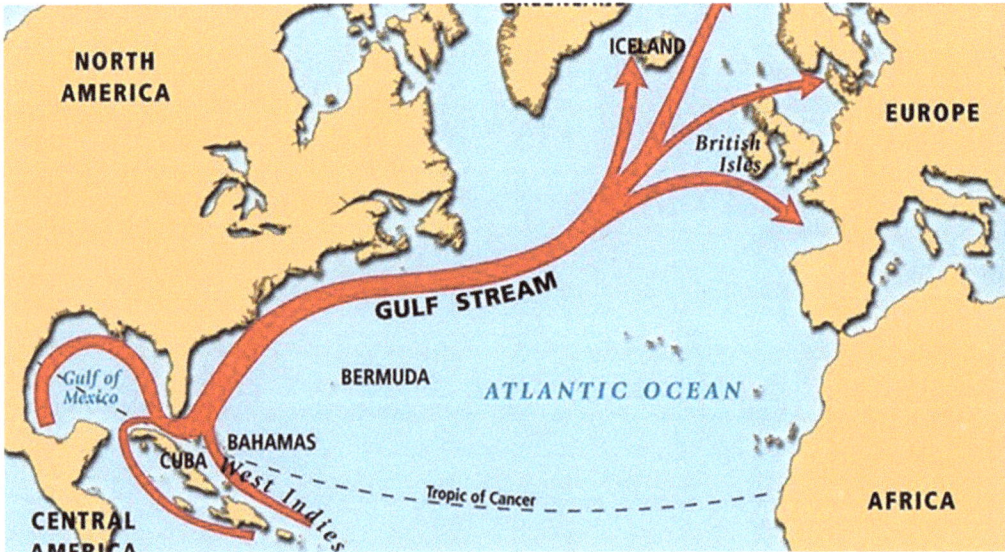

The Gulf Stream is made of water warm in the Gulf of Mexico that flows North through the Atlantic Ocean.

Surprisingly, only one percent of the Island is used for agriculture, but it significantly contributes to the economy. Agriculture ranks third after tourism and fishing to make up the national income. The southern part of the island entertains most of the geothermal tourism and contains the largest number of farms. The north has extensive areas of green acres supporting crop and livestock farming.

Physical map of Iceland. About 78% of Iceland is agriculturally unproductive. Green areas along the periphery of the island are where agriculture is possible.

Raising sheep (the traditional mainstay for generations of Icelandic farmers, and cattle make up the majority of livestock while pigs and poultry are also raised. Iceland is self-sufficient in the production of meat, dairy products, and eggs. There are over 700 dairy farms in Iceland.

A variety of hearty food crops, such as potatoes, turnips, carrots, cabbage, kale, and cauliflower are grown. In the more populated south, other subtropical crops such as tomatoes, cucumbers, green peppers, flowers, and potted plants are grown in greenhouses heated with geothermal energy. In some cases, artificial light is used to account for the shorter winter daylight hours.

Grass, a perennial crop occupies over 90 percent of the agricultural land, is the main fodder crop which grows well in the long periods of daylight in the short, cool summers. Being a perennial crop, it can start growing again as the weather warms. Rye and barley are also grown. Ninety nine percent of agricultural land is dedicated to growing fodder. The cool climate provides certain advantages for agriculture as the lack of insect pests means that the use of agro-chemical like insecticides and herbicides is low. There is a general lack of pollution from the sparse population which means that food is less contaminated with artificial chemicals.

Farm in Klofningsevegur, in northwest Ireland. Photo by Rick Bein July 11,2022

Traveling in the northwest of the island, I found a predominance of hay and sheep. I was constantly aware of sheep roaming freely, crossing the road with abandon. There were signs alerting the drivers of this hazard. Someone told me to be more careful after one sheep crosses, because she might soon be followed by her lamb(s). Fortunately, I had no such collision.

When asking farmers what crops they grow, I was expecting some variety, but the answer was "hay". "That's all?" "Our hay is grown to feed our sheep." "Any other animals?' I asked. "A few horses for rounding up the sheep when summer ends."

slandic horses (from Holmavik) and sheep came with the first settlers 1000 years ago and have become unique breeds as Islanders prevent any genetic mixing by preventing importation of different breeds. Photos by Rick Bein and Ellis Pawson, July 2022.

When asked why sheep are not confined in corrals, they answer "Only in the winter do we keep them in our pastures. In the summer we release them to graze the wild grasses that grow on the land which is too hilly for farming. The more level lands are needed for growing hay to feed the sheep during the winter. Hay is stored in those big plastic covered bales for that purpose".

After mowing the hay, it is stored in heavy plastic that keeps the hay preserved and free from mold and any additional moisture. This is a technique used around the world. Photo by Rick Bein July 10, 2022, Fljostshlioarvegur, Iceland.

Over the summer ewes and their newly born lambs wander freely and are scattered throughout the hills. As temperatures become cold, communities organize round ups using their horses to bring in the sheep. All the sheep are driven into a sorting corral (rittir) where each owner claims his sheep based on his cut marks in the ears of the ewes. The lambs stay close to their mothers and are collected at the same time. As the lambs have grown to market size, they are sent to the packing plant.

Rittir where sheep are sorted after bringing them out of the high ground during the Fall round-up. Photo taken in Fljotshlioarvegur by Rick Bein June 2022.

The ewes are then moved into the hay fields where the large bales of hay remain. These are opened one at a time as the sheep consume them over the winter. Rams begin breeding and the ewes and the starts over cycle again.

Hardy vegetables are grown in small plots near the households throughout the island. These root vegetables include potatoes, beets, carrots, and rutabagas, growing underground are less affected by the cold weather can be harvested later. Other cold weather vegetables such as cabbage and rhubarb are grown outside, but do better in green houses.

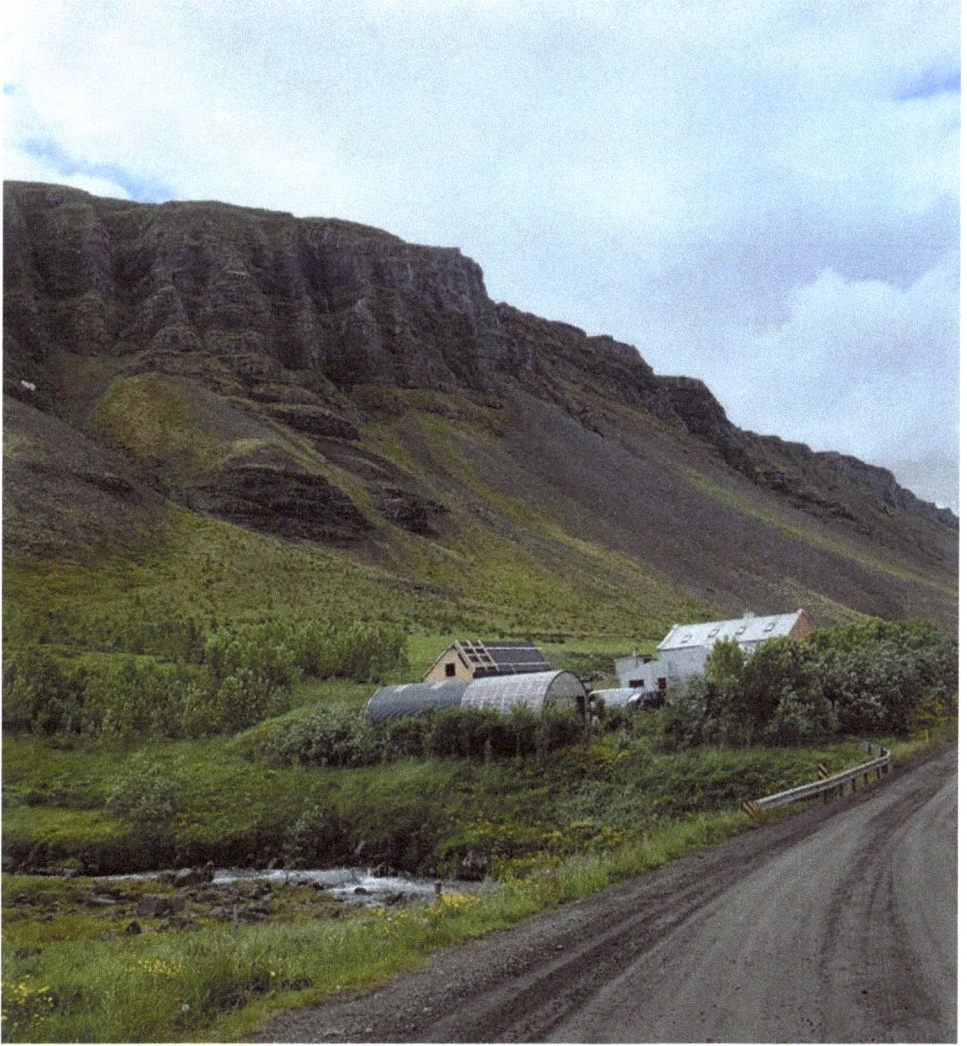

This Farmer keeps a greenhouse to enhance his family and neighbors' diets. Klofningsvegur 371 photo by Rick Bein July 10. 2022.

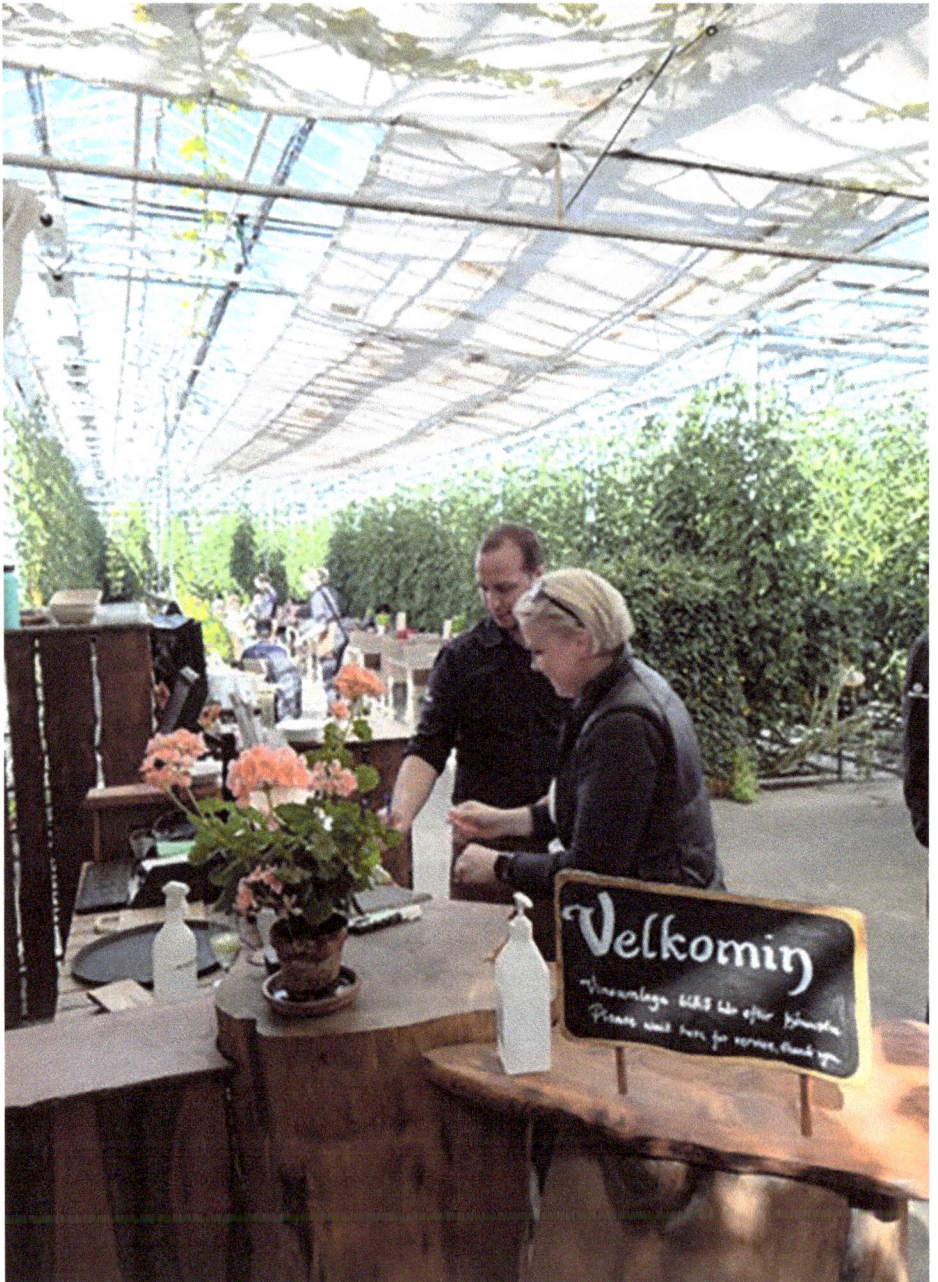

Greenhouses supply most of the fresh vegetable to Reykjavik and other urban centers in the south, Tomatoes are grown in this high-tech green house where they are tied vertically to twelve-foot cords attached to the ceiling. Photo taken in Lynbrau, Reykholt by Rick Bein July 2022.

There are some challenges to Icelandic agriculture, such as the threat of invasive species from outside the country. Alaskan Lupin is an example that seems to be taking over un-vegetated land and invading farms. Some think it helps control soil erosion. Most feel that the jury is still out about this plant and do not like it.

Icelanders have been careful to preserve the purity of their sheep and horse breeds by forbidding the importation of any foreign breeds.

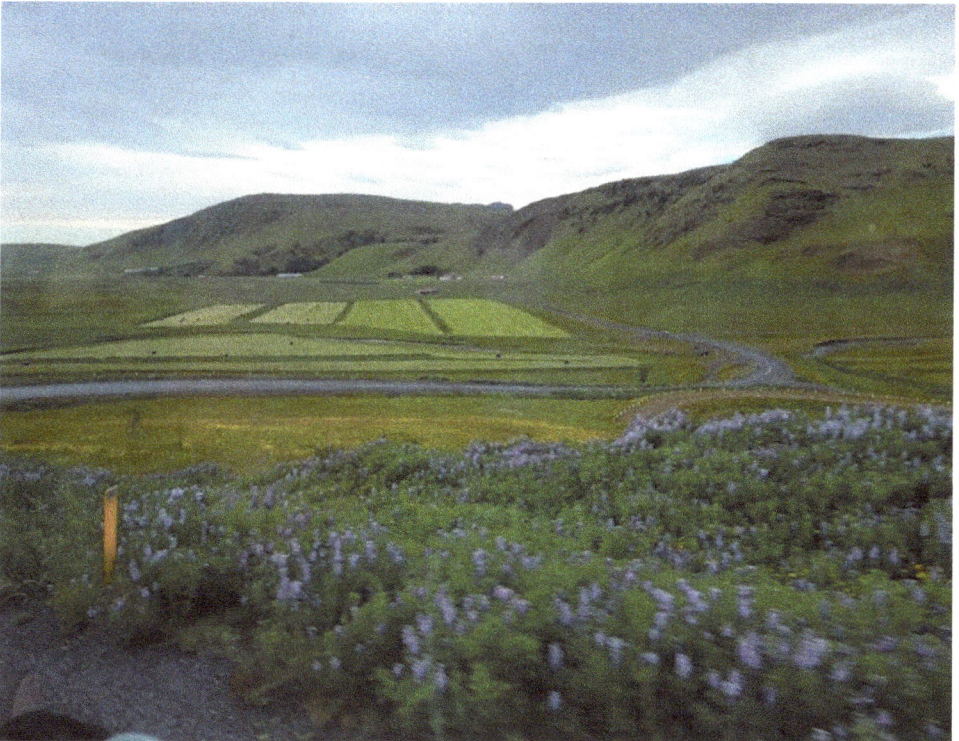

This manicured farm in the background is facing the spread of the Alaskan Lupin in the foreground. Photo by Rick Bein July 3, 2022. Location 1,871, Iceland.

References

Data was collected during the June-July 2022 National Council for Geographic summer teacher field course to Iceland.

Google Earth, https://www.google.co.jp/intl/ja/earth

Farms of Iceland: Names, Facts, and Features – Iceland.org   Erpsstadir Farm

www.iceland.org/geography/farm

Encyclopedia Britannica https://www.britannica.com/place/Atlantic-Ocean

Encyclopedia of the Nations "Iceland – Agriculture" www.iceland.org/geography/farm

Iceland Review – Iceland Review https://www.icelandreview.com

Iceland Weather, Climate, & Temperature Year-Round   https://guidetoiceland.is/travel-info/climate. Author: Nanna Gunnarsdóttir

NOTES

Rick is indebted to those who have been instrumental in supporting this effort by editing, photographing, remembering details and suggestions. These are Jeanie Bein, Alex Bein, Linda Beach, Chris Lett, Maryellen Bein, Candy Riggins, Mary Bein, Glbert Nduru, Mohamid Ibrahim, Elizabeth Lynch, Kristen Hamid, Randy Beach, Molly Funk, and Robert Melvin. I hope you have enjoyed my writings.

If you wish to explore more, feel free to search the following title.

Traveling farmer by Rick Bein

Rick Bein

DR. RICK BEIN

www.ingramcontent.com/pod-product-compliance
Lightning Source LLC
Chambersburg PA
CBHW052112030426
42335CB00025B/2948